ROCKHURST COLLEGE LIBRARY

0 0006 0075506 0

D0002829

ON THINKING STATISTICALLY

Also from the Administrative Staff College
and published by Heinemann

New Thinking in Management: F. de P. Hanika

Morris Brodie

ON THINKING STATISTICALLY

A Short Introduction

NEW AND REVISED EDITION

Distributed in the United States by
CRANE, RUSSAK & COMPANY, INC.
347 Madison Avenue
New York, New York 10017

HEINEMANN : LONDON

Published in co-operation with
the Administrative Staff College, Henley

William Heinemann Ltd

15 Queen St, Mayfair, London W1X 8BE

LONDON MELBOURNE TORONTO

JOHANNESBURG AUCKLAND

© The Administrative Staff College 1963, 1972

Originally published by Hutchinson & Co. (Publishers) Ltd. 1963
Reprinted 1966, 1969, 1970
This second edition published 1972

434 90185 7

Printed in Great Britain by
Willmer Brothers Limited, Birkenhead

HA
29
B8282
1972

5.50

July 1976

Emery – Pratt

To Judith and Deborah

91931
ROCKHURST COLLEGE LIBRARY

ROCHFORD COLLEGE LIBRARY

'You are oversimplifying,' I told him, 'but it is refreshing in an age of overcomplication.'

JAMES THURBER

Preface to the Second Edition

On Thinking Statistically was published to meet a particular need. In this, it has met with some success. It has been in steady demand, reprinted on three occasions, and paid the compliment of being pirated. It was originally written for managers. In practice it has met a wider need of busy people in all walks of life – in business, commerce, the public service, and the professions – looking for a succinct layman's introduction to statistics and an explanation of the help they can get from a statistician. In addition it has been used in universities, technical colleges, and schools to give students a non-specialist introduction to the subject. I am grateful to a number of people, managers and others, who have been in touch with me, for their helpful comments and suggestions.

Some of the suggestions I have acted upon, in preparing this revised edition. One, made not infrequently – that I should substantially lengthen it – I have resisted. My own agonies bear out the experience of other writers that a short book does indeed take much longer to write than a long one, but finally the result is more satisfying and would seem to me to be in the interests of the kind of reader for whom this book is intended.

For the non-specialist reader who wants to go into the subject more fully, there is now a better choice than was the case a few years ago. As for the reader who prefers to buy his books by weight, he is already well catered for. Nevertheless, *On Thinking Statistically* has grown a bit. I have expanded the original Preface and made it now, more appropriately, the first chapter. Other changes of some substance – I hope for the better – have been made in Chapters 2 and 5. I have also added a chapter, on 'Using the Experts', because something needed to be said on that.

While the focus of my own professional work over the years has widened, I remain convinced that it is important to help managers at all levels to look at their problems with some statistical understanding. I am no less convinced that the potential for exploiting even the most basic statistical notions and the simplest statistical techniques remains vast. I hope that the revised edition will continue to serve this purpose.

Once more I was glad to have the help of Mrs P. Hayes, who dealt so ably with much of the typing and preparation of the manuscript for publication; this work was admirably rounded off by Mrs Gay Swadling.

I am indebted to my colleagues Mr Andrew Life and Mr Philip Montagnon for reading and commenting on this revised edition; to be able to turn to friendly but frank critics is indeed a privilege. The responsibility for shortcomings remains mine alone.

MORRIS BRODIE

Contents

Preface   ix

1. Thinking Statistically   1

2. Collecting, Assessing, and Systematizing Data   9

   The quality and sources of data – Accuracy –
   Comparisons – Index numbers – Classifying and
   grouping data – 'On the average' – Measuring
   the scatter of data

3. Taking Samples   31

   Sampling and its risks – The basis of reliable
   sampling – The 'normal' distribution curve –
   The quality control chart – Quota sampling

4. Analysing Trends   47

   Forward thinking and forecasting – 'Time series'
   analysis – Graphing – Logarithmic charts –
   Moving averages and measures of seasonal effect –
   Correlation – The limits of forecasting

5. Handling Complexity   65

   Model building – Quantifying problems which
   resist measurement – Notions of self-regulating
   systems – Methods of handling, storing, and
   processing data – The more analytical approach
   to fact-finding and diagnosis

6. Using the Experts   77

   A Note on Further Reading   86

   Acknowledgments   87

   Index   89

Chapter 1

Thinking Statistically

'A mind all logic is like a knife all blade,
It makes the hand bleed that uses it!'

Tagore*

*Collected Poems and Plays of Rabindranath
Tagore*, Macmillan, 1950, page 312

I

Thinking Statistically

'... have you ever noticed that those who have a natural capacity for calculation are, generally speaking, naturally quick at all kinds of study; while men of slow intellect, if they are trained and exercised in arithmetic, if they get nothing else from it, at least all improve and become sharper than they were before?'*

'Statistical ignorance and statistical fallacies are quite widespread and quite as dangerous as the logical fallacies which come under the heading of illiteracy. The man who is innumerate is cut off from understanding some of the relatively new ways in which the human mind is now most busily at work. Numeracy has come to be an indispensable tool to the understanding and mastery of all phcnomena, and not only of those in the relatively close field of the traditional natural sciences. The ways in which we think, marshal our evidence and formulate our arguments in every field today is influenced by techniques first applied in science. The educated man, therefore, needs to be numerate as well as literate.'†

From Plato to Crowther is a span of almost 2,400 years, but an equally long history is to be traced behind not a few con-

---

* Plato, *The Republic*, 526. Everyman's Library, Dent, 1942.
† *15 to 18: A Report of the Central Advisory Council for Education.* Crowther Report, para. 401. HMSO, 1959.

temporary educational problems. Plato was not concerned with a training in arithmetic for the skills of calculation. They were the least part of it. For him, mathematics was an instrument for training in logical thought, a way of sharpening the intellect.

Statistics may be said to justify a place in education not only for the expertise it represents, important though that is, but also for the particular nature of statistical reasoning, with its call for a right balance between the deployment of technique and the exercise of judgement. This balance was less of a problem in olden times, when the study of statistics was understood to be the study of matters of state.

A statist, regrettably an archaic term today, was primarily one skilled in state affairs, only secondarily one who dealt with statistics. In 1791, Sir John Sinclair, introducing *The Statistical Account of Scotland* – in fact the first of what was to be a series – hoped that it would enable the reader '... to form some general idea of the State of the Kingdom. ...' He also quoted 'a respectable citizen' who commended the work in these terms: 'That no publication of equal information and curiosity has appeared in Great Britain since Dooms-day Book; and that, from the ample and authentic facts which it records, it must be resorted to by every future Statesman, Philosopher, and Divine, as the best basis that has ever yet appeared for political speculation.'

It was this link with matters of state which caused a lot of trouble in days gone by, when proposals were mooted from time to time for collecting data. It was the need for reliable population figures, for example, which led to a Bill being introduced in March 1753 for an annual census. It was strongly opposed and finally rejected. Much of the opposition was on grounds of religion; there were fears that numbering the people would be followed 'by some great public misfortune or epidemical distemper'.

The Manchester Statistical Society, founded in 1833, one year before the London Society, defined its objects as 'The collection of facts illustrative of the condition of Society, and the discussion of subjects of Social and Political Economy ... and the dissemination as widely as practicable for the benefit of the community of the results of such collection and discussion.'

How then did the contemporary, more limited, view, which equates statistics with just dull figures, with a set of techniques to process, manipulate and analyse quantitative data, take hold?

In part, it has to be acknowledged that many of the older generation have good reason for taking a jaundiced view of anything to do with mathematics. Too many of them learned their mathematics through tedious bath-filling-and-emptying exercises. The Hindus showed much more imagination in the exercises they were setting, as long ago as the eighth century, for teaching algebra:

'A necklace was broken during an amorous struggle. One-third of the pearls fell to the ground, one-fifth stayed on the couch, one-sixth was found by the girl, and one-tenth recovered by her lover; six pearls remained on the string. Say of how many pearls the necklace was composed.'*

Exercises of that kind might have induced a somewhat more receptive attitude towards mathematics than is common today amongst adults.

Paradoxically, the antipathy to statistics coincides with the important analytical developments which recent decades have witnessed and which have perhaps led to a preoccupation with techniques.

Something of a corrective may be under way, associated with the interest that is now developing in the application of statistical analysis to decision-making and forward planning.

'Statistics is concerned with decision-making in the face of uncertainty' is now a frequently used definition, which has the merit which many traditional text-book approaches lacked, of relating method to purpose.

In applying science to uncertainty and using dispassionate analysis to attack seemingly intractable problems, there is inspiration to be drawn from the pioneers. Galton, in a memoir

* Quoted in Tobias Dantzig, *Number: The Language of Science*. George Allen and Unwin Ltd., 1930, page 82.

entitled *Statistical Inquiries into the Efficacy of Prayer*,* which he wrote in 1872, observed that he could see no reason why the efficacy of prayer should not be 'a perfectly appropriate and legitimate subject of scientific enquiry', and he proceeded to show how this might be approached. Do statistics show that those who pray for recovery from disease tend to recover more rapidly than those who do not? Do those whose lives are prayed for more than other classes of the population – the Royal Family, clergymen, for example – live longer than others? Do the insurance companies charge less for missionary vessels than for ordinary trading vessels, since presumably they enjoy greater Divine protection?

Galton's message, that the scope for scientific enquiry and reasoning is much greater than conventional thinking allows, holds good no less today. Statistical analysis and statistical reasoning have proved to be powerful weapons in the armoury of rational enquiry in the sciences and in the arts. The fact that statistical techniques have found application in such diverse fields as literature, archaeology, medicine, and business should be enough to satisfy any manager on that point, even the one who prefers to regard himself more as an artist than a scientist. Either way, he is someone who puts his knowledge, skills, and judgement to the test of action and commitment, and the statistical approach is action- and decision-centred.

Regrettably, there are still those, even in high places, who are contemptuous of figures and mutter all too readily 'Lies, damned lies and statistics'. Statistics do not lie. People do – and if that is what is meant, it is better stated bluntly, not deviously. Usually the situation is less dramatic. People use figures misguidedly usually through innocence or ignorance. The man-in-the-street may be excused, given the examples he gets daily. The manager cannot reasonably plead that he is innocent or ignorant. He must be prepared to do something about it. He must try to know enough about statistics to put them to good use as well as to ask the right questions about them.

* Sir Francis Galton, Three Memoirs printed for the Eugenics Society, December 1951. Richard Clay & Co. Ltd., Suffolk.

6

Managers need an entrée into the world of the statistician rather than instruction in specific techniques and skills. The attempt to impart insight rather than skills involves two important assumptions which are worth making explicit. The first is that the general notions of a statistical approach to affairs can in fact be usefully understood without having to acquire the specialist expertise. The soundness of this assumption is not solely a function of the abilities of the non-specialist. It also depends upon the skills of the teacher, upon whether or not he can make the notions of a specialist discipline understandable to the lay mind. The second assumption is that an administrator or manager armed with such notions will be the better able to recognize the kind of problem amenable to specialist statistical treatment. The difficulty here is in establishing links between problems familiar to the practitioner and the peculiar skills which the statistician might bring to bear upon them.

The chapters which follow seek to illustrate why, with the ever-widening use of figures and statistical notions and the impressive elaboration of statistical techniques of the last decades, those in positions of responsibility in all walks of life need to think more statistically.

Chapter 2

Collecting, Assessing, and Systematizing Data

'The gretteste clerkes been noght the wysest men'

Chaucer*

*The Reves Tale, in The Poems of Geoffrey Chaucer. Oxford University Press, 1962, page 470

2

Collecting, Assessing, and Systematizing Data

It is important to judge the quality of data which are the raw material of statistical analysis.

Any undertaking generates a considerable amount of data internally and assessment of accuracy and reliability should be relatively straightforward. This is not to say that such data will necessarily be particularly accurate, but rather to make the point that there should be no difficulty about establishing what order of accuracy to attribute to such data, or about ensuring that this is known to the users.

Practice usually belies this. Figures are disseminated in many undertakings with rarely an indication of how they have been compiled or of the degree of accuracy governing them, and as a result there can be a great deal of futile discussion.

Definition of terms and of units requires care. Does a figure of monthly sales mean orders received, or goods despatched, or goods invoiced? A good unit for statistical purposes will be easily identifiable, capable of being measured satisfactorily, and stable. How do the metre, the £, the 'man in the street' all widely used units – stand up to these criteria? How hard are the figures expressed in 'hard cash'?

When data are obtained from outside sources, some knowledge of the way the data have been collected, compiled and presented is equally essential. The Central Statistical Office set a good example by publishing an annual Supplement to the *Monthly Digest of Statistics* which gives definitions and explanatory notes.

11

In this context, a word is appropriate on the diverse range of published statistics.

Most government departments have statistical sections and some publish departmentally. The main source for the public, however, is the Central Statistical Office whose functions are described in a useful booklet, *Government Statistical Services*. A glance at an HMSO catalogue would quickly illustrate the very wide range covered.

For initial and general reference, the *Monthly Digest of Statistics* is particularly useful. At present (December 1971) it is divided into nineteen major sections:

1. National Income and Expenditure
2. Population and Vital Statistics
3. Labour
4. Social Services
5. Agriculture and Food
6. Production, Output and Costs
7. Fuel and Power
8. Chemicals
9. Metals, Engineering and Vehicles
10. Textiles and other manufactures
11. Construction
12. Retailing and Catering
13. Transport
14. External Trade
15. Overseas Finance
16. Home Finance
17. Wages and Prices
18. Entertainment
19. Weather

Earlier comparative data are normally included, e.g.

Consumers' expenditure: annually from 1965, quarterly from 1969.
Rainfall: average for the periods 1916–50, monthly from January 1969.

Unemployment: annually from 1965, monthly from October 1968.

The *Annual Abstract of Statistics, Economic Trends, Board of Trade Journal, Employment and Productivity Gazette* and the monthly *Economic Progress Reports* and *Assessments* prepared by the Treasury Information Division are also well worth knowing. Finding one's way through the complexities of published statistics is not always easy; B. Edwards, *Sources of Economic and Business Statistics* (Heinemann, 1972) provides a valuable up-to-date guide.

The development of official statistics has taken big strides in recent years, particularly following the Estimates Committee Report of December 1966 on Government Statistical Services.

One of the clearest indications is the decision taken by the Central Statistical Office to publish a regular bulletin on developments in official statistics, called *Statistical News*. The first number came out in May 1968. Anyone working with statistics would find this bulletin invaluable in keeping informed of changes and developments, particularly as improvement is a continuous process.

In the first issue, the Director of the Central Statistical Office, Professor Moser, explained that the main changes are intended to produce a more integrated system of official statistics, and to improve the usefulness of the statistics to Government and to other interests, notably the business world, trade unions, economists and the Press.

These developments entail new responsibilities for the Central Statistical Office. This is especially so in managing, co-ordinating, and integrating the statistical system, assessing important technical developments such as the use of computers, and improving standards. No less important, the need is acknowledged to improve the dissemination of official statistics and there have already been a number of excellent innovations. For example, a booklet *Government Statistics for Industry* is available which sets out the type of business and economic information available, the main published sources and telephone numbers for further details. A new Business Statistics

Office has been set up, whose tasks include compiling a Central Register of Businesses, building up data banks, and introducing an improved system of industrial statistics.

All this is to the good, yet the problems which face policy makers in being guided by statistics are severe. For example, for some years the American Government was very much influenced in its economic policy by the gap estimated to exist between potential and actual output, as measured by gross national product. Following a period when the figures had shown a persistent gap, the Government took expansionary measures. In point of fact, the gap proved to be statistical, though the statisticians needed time to improve and revise their data before they could explain it. Meanwhile, of course, the Government had to battle with the effects of its expansionary measures on what was in real terms an already fully employed economy, although the statistics at the time showed otherwise.

In the case of officially published statistics, we tend to be mostly concerned with those having to do with the economy. Currently, this is no doubt regarded as the most critical aspect of the policies of Government. However, even assuming a highly developed, effective and timely apparatus of economic data, there remain other important statistical needs which are being met hardly at all. Governments increasingly legislate for social as well as economic progress; think, for example, of the scale of investment in education and health. The main available data on this kind of investment are in terms of input – the resources being put into the various activities. By contrast data in terms of output – the purposes for which the inputs are made – are woefully lacking and increasing attention will need to be given to this, as well as to data for planning and control, for measuring performance and for prediction.

Many international organizations and agencies publish their own comprehensive statistics. Two examples: the *Bulletin of Labour Statistics*, published quarterly with intervening supplements by the International Labour Office, giving international comparative data on employment, hours of work, ages, and consumer prices: *International Financial Statistics*, published monthly by the International Monetary Fund, containing nation-by-nation summaries and world tables of data vital to international trade. Of growing importance are the efforts being made to standardize the collection and presentation of statistics for better international comparison. Special international studies in methods are under way and some progress is being made in this difficult task.

Many journals and newspapers regularly give selected published statistics, and offer special reviews from time to time. The *Financial Times*, the *National Institute Economic Review*, bank journals, all repay scrutiny.

In both the quality and range of published statistics there is constant change and improvement and any summary would be quickly dated. The financial columns of the Press and several periodicals and journals normally draw attention to significant developments.

More data are available than is usually realized and ignorance of sources leads to much waste of effort. In establishing what data are available in a given field, such services as Aslib (the Association of Special Libraries and Information Bureaux) renders are particularly helpful.

ACCURACY

Accuracy is relative since measurement involves approximation and one must beware of a spurious appearance of accuracy.

It is not sufficiently recognized that figures as they stand give an impression of accuracy. Consider the figures given when six study groups were asked to indicate the number of words in their reports:

A: 3,500	D: 3,320
B: 3,250	E: 2,688
C: 3,460	F: 2,250

Looking at these figures, it would not be entirely unreasonable to infer that A estimated to the nearest 500 words. It is

15

possible that every single word was counted and the total in fact came to just 3,500. Without knowing this, however, one might be forgiven for presuming an accuracy of ± 250 words. On the same analogy, B and F may have counted to the nearest 250 – or was it to the nearest 50? – C and D perhaps to the nearest ten or twenty. E presumably counted every single word.

The only way of insuring against such presumptions is to give the reader an idea of the limits of reliability to attach to figures, a habit which should be regarded as a hallmark of good practice, though it is all too rare.

COMPARISONS

We often wish to measure change, which we do by comparison. Comparisons can be drawn by the use of percentages and ratios, but the very simplicity of such calculations can conceal how misleading they can be. They need handling and judging with caution.

Take the case of the executive on £3,000 a year, invited by his Managing Director to accept a 50% cut for the next twelve months because the firm is in some difficulty. He resigns himself with good grace to his £3,000 − 50% = £1,500. Things get better. His 50% is then restored; £1,500 + 50% = £2,250. The perverse ways in which percentages can be calculated and exploited in argument are numerous and plentifully illustrated in the daily Press.

Consider the figures of a particular by-election:

	% Votes cast	% Total electorate	Compared with General Election	Actual votes
Conservative	57	41	Drop 11%	14,000
Liberal	20	14	—	4,800
Labour	16	11	Drop 2%	3,800
Scottish Nationalist	7	5	Drop 8%	1,800
				24,400
			Total electorate:	34,000

16

How should we read these figures? The answer depends on who is doing the reading. The Conservative candidate can claim that he won with a majority vote and it is true. He got 57% of the votes cast. The defeated candidates can claim that the Conservative won on a minority vote. That too is true, because he got only 41% of the total number of possible votes. Labour can claim that it came out best relatively, because it registered the smallest drop compared with the pattern at the time of the general election. All candidates should quote figures in relation to votes cast, not total electorate, because that makes them look higher. None of these different uses of the same figures is incorrect, but each has a different impact – which is why the ordinary person needs to be very wary indeed about the way figures are bandied about.

One more example. A consultancy firm took a full page in a national newspaper to advertise that its turnover had increased in year 2 over year 1 by 25%. Figures were then given for the first 24 weeks of year 3 and these were compared with the first 24 weeks of year 2 – showing a growth of 43%. Implication: a very fast rate of growth. However, taking the same data, we can come to a different conclusion. Instead of taking the first 24 weeks of year 2, let us take the last 24 weeks, which it is possible to estimate without much difficulty. If we then compare the first 24 weeks of year 3 with the last 24 weeks of year 2, the growth rate is only 10% and it could be argued that this calculation is much more significant, because it relates to the more recent past.

	Turnover	
Year 1	£4·5m	
Year 2	5·6	+25%
First 24 weeks of year 2	2·2	
First 24 weeks of year 3	3·2	+43%
Last 24 weeks of year 2	2·9	
First 24 weeks of year 3	3·2	+10%

Then the interpretation could be that this was a business which had been growing at a rate of 25% per annum and now this rate has fallen to 10%. Far from growing spectacularly the business was really on the decline and – as is often the case – the firm was not really aware of this. Apart from the statistical lessons we can draw from playing with figures in this way, perhaps this example also suggests that even consultants may need consultants, to help them to understand better what is happening inside their own business.

In a given year the death rate in Henley, at 17 per 1,000 of the population, was almost twice the death rate for industrial Bootle, which stood at 9·9 per 1,000. Bournemouth's at 16·3 per 1,000 was significantly higher than that of the Rhondda at 14·3 per 1,000. What conclusion should we draw, if we aspire to longevity? Live in Bootle rather than in Henley, in the Rhondda rather than in Bournemouth? In the same year, taking the number of men who died in England and Wales in the age group 45–49, for every one who was single, six were married. (Actual figures 1,199 single and 6,839 married.) Is marriage that much more lethal?

INDEX NUMBERS

Index numbers measure changes over time in the values attached to such slippery notions as the cost of living, in an attempt to give them some kind of firmness. The elements of which they are made up may be numerous and in practice may not easily lend themselves to direct measurement, yet, cautiously used, such indicators of relative change can be of considerable value. Index numbers are frequently met in economic studies. They are finding growing application in business, for example as indicators of price movements or as a basis for wage agreements.

Figures of expenditure on imports, and earnings from exports provide a convenient illustration of their use. The tables* give the figures for the years 1962–70, together with unit value and volume indices.

* *Annual Abstract of Statistics*, HMSO, 1971. Tables 269, 270, 271.

18

Imports and exports 1962–70

	Imports £ million	Exports £ million
1962	4,628	4,062
3	4,983	4,365
4	5,696	4,565
5	5,751	4,901
6	5,949	5,255
7	6,437	5,230
8	7,898	6,434
9	8,315	7,339
1970	9,052	8,063

Unit value and Volume index numbers
1961 = 100

	Imports		Exports		Terms of trade*
	Unit value	Volume	Unit value	Volume	
1962	99	103	101	102	102
3	103	107	104	108	101
4	107	119	106	111	99
5	107	120	109	116	102
6	109	122	113	121	104
7	109	132	114	119	105
8	121	146	123	136	102
9	126	149	127	150	101
1970	132	157	136	155	104

* Export unit value index as a percentage of the import unit value index.

Compare the year 1964 with 1963. Imports rose by £700 million, or some 16%. How much of this rise was real and how much due to rising prices? The unit value index shows a rise of 4 points, which indicates that a good part of the rise in the total import bill was to pay for a real increase, a bigger volume of imports. This shows up in the volume index; it rose 12 points. How did exports do? In 1964 they grew by £200 million – by about 5% – over 1963. Export prices went up 2 points. Some part of the growth of exports was due to an increase in volume; this shows up in the rise of 3 points in that index.

A further indicator, known as the terms of trade, relates

the price movements to each other, reflecting the way export prices have moved in relation to import prices. Over the two years, imports rose in price relatively more than exports. Put another way, the terms of trade went against us, which is reflected in the drop of two points in that index.

In the same way, we can look at the rise of £1,460 million in our import bill in 1968, compared with 1967. We see from the index numbers of unit value and volume that this was made up in part of a very sharp rise in import prices, in part of a rise in volume. Exports rose by £1,200 million. Some of this is explained by the 9 points rise in prices, the rest by the 17 points rise in volume. Again, we had to pay relatively higher prices for our imports than we were able to get for our exports – the terms of trade moved three points against us (though over this nine year period, on balance, the terms of trade if anything tended to be in our favour).

Establishing an index which attempts to pin down, in the form of a measurement, such slippery notions as the cost of living or the basis for a minimum wage is by no means easy. Some kind of yardstick is first necessary. The difficulties of establishing an acceptable yardstick show up vividly in the case of a region such as West Africa. 'Under West African conditions it is difficult to calculate the cost of a subsistence diet, let alone a minimum subsistence wage. The diets of people of different regions vary a great deal. The price of farm products depends largely on the cost of transport, marketing and storage. The price of fish on the coast may be one-quarter of the price in the interior. Plantain or yams cost far less near the growing areas than in the town markets. An index figure based on a subsistence diet at the coast could be quite deceptive inland. The inefficiency of transport, of storage and of preservation, exaggerates the seasonal variations of supply, and results in greater variations in the cost of a given diet. It has, therefore, to be admitted that the difficulties of establishing a minimum wage are immense.'*

Complex problems have to be resolved in determining the best basis for the calculation of such an index. Nevertheless,

* J. L. Roper, *Labour Problems in West Africa*, pages 90–1. Penguin, 1958.

despite the limitations and imperfections of the result, this may constitute the only practical way of putting some kind of measure on the cost of living and of comparing changes over time.

The main features of construction of an index can be seen in the way the index of retail prices in this country is calculated. The *Industrial Relations Handbook* (HMSO, 1961, pp. 186–8) briefly explains it, while conclusions of the Cost of Living Advisory Committee on how the index might be improved are contained in Cmd. 1657 of March 1962, 'Report on Revision of the Index of Retail Prices'.

Features of particular importance in index calculation are:

1. Choice of BASE PERIOD. The movement can easily be damped down or exaggerated, depending upon the base period chosen. A period of stability is clearly to be preferred. Often the average over a period is used.
2. The ITEMS to be incorporated. These must provide a good coverage and must be unambiguously definable. Where sampling is involved, the methods must be rigorous.
3. The WEIGHTS allocated to items. These reflect the relative importance of individual items within the group.
4. The METHOD OF CALCULATION. There is no such thing as one right index and there are many possible methods of calculation. Skill and experience are necessary to evolve index numbers appropriate to purpose, while their use and interpretation need care.

CLASSIFYING AND GROUPING DATA

A mass of original data may provide the essential raw material for subsequent analysis, but too much detail can be a nuisance. It is necessary, therefore, to bring some order into the data. This is done by classifying, grouping, arranging in summary form, and by presenting the now more systematically organized data in tables, graphs and charts. There are canons of good practice on how this should be done. There are daily

21

examples of bad or questionable practice. Most introductory text-books on statistics go into this, some more entertainingly than others.

The graph of sales of Sunday newspapers in the United Kingdom is an example of the illustrative value graphs can have – and of the caution needed in interpreting them (see Figure 1). The time scale is long enough for fairly definite trends to emerge and for us to be able to see the changing pattern in some perspective. Nothing can disguise the long-term decline of the *News of the World*, though no other newspaper has overtaken it. We note the small but reasonably steady growth in sales of the serious Sundays. However, we could have a more favourable impression of the situation for the other popular Sundays than the facts warrant, if we are not also aware of the fact that during this period there were popular Sundays which ceased publication or were absorbed by others.

By grouping data, it is easier to see how variable the data are. A FREQUENCY DISTRIBUTION shows the way in which data are distributed against readings.

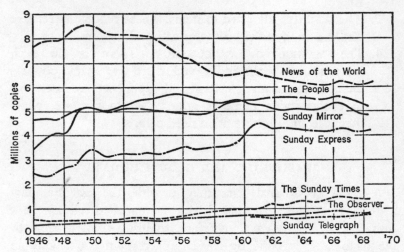

Figure 1. Sales of Sunday newspapers: United Kingdom 1946-69 (From *Britain in Figures: A Handbook of Social Statistics*, A. F. Sillitoe, Penguin, 1971, page 181; source of data Audit Bureau of Circulations.)

Take the case of trade unions as an example. If one were enquiring into the structure of the trade union movement in the United Kingdom in 1970, one could start with the individual membership of each of the 481 unions, in total a cumbersome mass of data. It would be a tedious but necessary preliminary to rank the figures in some order of magnitude. By grouping the unions in the form illustrated in the table,* the mass of original data is reduced to manageable proportions.

Size of Trade Unions: 1970

Number of members		Number of unions	% of total number of all unions	Total member- ship (a)	% of total membership of all unions
Under 100		90	18·7	4,000	0·0
100 and under	500	116	24·1	29,000	0·3
500 and under	1,000	50	10·4	35,000	0·3
1,000 and under	2,500	59	12·3	99,000	0·9
2,500 and under	5,000	50	10·4	172,000	1·6
5,000 and under	10,000	30	6·2	199,000	1·8
10,000 and under	15,000	13	2·7	155,000	1·4
15,000 and under	25,000	21	4·4	403,000	3·7
25,000 and under	50,000	13	2·7	452,000	4·1
50,000 and under	100,000	16	3·3	1,111,000	10·1
100,000 and under	250,000	14	2·9	2,188,000	19·9
250,000 and more		9	1·9	6,155,000	55·9
Totals		481	100·0	11,000,000	100·0

(a) The figures have been rounded to the nearest 1,000 members with the result that the sum of the constituent items does not agree with the total shown.

There is a further obvious advantage in doing this. One can quickly pick out certain impressive facts, such as the great concentration of membership in a very few trade unions. One can infer something about the organizational problems of the TUC in dealing with the 33,000 trade unionists scattered amongst the 206 unions all of which have less than 500 members.

'ON THE AVERAGE'

It is often convenient to take an average reading and regard it as typical of the whole. But the notion of an average is loose.

* *Department of Employment Gazette*, November 1971, page 1022.

To go back to the trade union data, how would one answer the question: What is the membership of the average trade union? Does the questioner mean the arithmetic average? Or the one that happens to be in the middle, if all the unions are listed in order of magnitude? Or the typical, however defined? The answers are different, depending on which notion is selected.

To avoid confusion, an agreed language is necessary. By way of illustration, the statistician would define:

(*a*) the familiar, arithmetic average as the   ARITHMETIC MEAN

(*b*) the most frequently occurring reading as the MODE

(*c*) the central reading when readings are ranked in order of magnitude as the   MEDIAN

For the trade union data, the arithmetic average – or mean – is roughly 23,000. The most frequently occurring size – or mode – is in the bracket 100–500. The central reading – or median – is somewhere between 500 and 1,000.

An average on its own really tells very little. Indeed it may confuse rather than inform. There is a salutary lesson from the early days of the Independent Television Authority's policy on advertising time.

Paragraph 2 of the Second Schedule of the Television Act 1954 reads:

'The amount of time given to advertising in the programmes shall not be so great as to detract from the value of the programmes as a medium of entertainment, instruction and information.'

In order to give more precision to this general obligation, the Independent Television Authority decided that it would work to the rule that the amount of advertising should not exceed six minutes an hour 'averaged over the day's broadcasting'. What happened subsequently disturbed some people. It prompted Mr Christopher Mayhew to introduce a Private Member's Bill on 12 November 1958 (which did not get very far) to limit more strictly the amount of time allowed to advertising. He maintained that:

(i) Evening-out the six-minute average over the day had not been envisaged by Parliament when it passed the Act.

(ii) In a typical week the average advertising time between 7 p.m. and 10 p.m. had been 9 minutes an hour. On three occasions there had been hours in which advertising had taken up over 20 minutes.

(iii) By concentrating advertising in peak hours programme contractors were making vast profits by this 'most public piece of systematic looting in British history'.

There may have been some ground for complaint. The ITA, in its annual report and accounts for 1958–9, stated that it had always kept under review the spacing of advertising over the day. It did find, however, that '... while the intention of the\ arrangements was being broadly achieved, particular programme patterns could lead to an over-concentration of advertising in particular hours. Accordingly, further modifications were introduced to prevent more than a tiny proportion of "clock" hours carrying more than eight minutes of "spot" advertising, and these were at the end of the year being progressively brought into force by the programme companies.'

What's in an average? In this case, it might be argued, a sum of £2 million, one estimate of the amount the Independent Television Authority gained as a result of this looseness of wording. It is worth pondering that neither the situation nor the controversy which followed would have arisen if the rule had been worded slightly differently, not just 'six minutes an hour averaged over the day' but 'six minutes an hour averaged over the day with a maximum of eight minutes'.

This example points to the importance of knowing not just the average figure for a bunch of data but also of having some idea of how scattered the data are around that average.

Taking the hour-by-hour record of advertising time, if one were told that it averaged six minutes per hour and varied from five to eight minutes per hour, this would present a very different picture compared with averaging six minutes per hour over the day, but varying from three to twenty. This information for each case – the average plus the two extremes – would be quite illuminating in telling us something about the impact of advertising upon the general structure of television programmes in the two different situations. An average on its own would have told us little.

MEASURING THE SCATTER OF DATA

It is a very useful thing, therefore, to devise measures which will indicate the scatter, variation, dispersion – terms which mean the same thing – of data around an average. There are a number of such measures, each having its own usefulness and limitations. A brief word on the most common.

RANGE

This is the difference between the largest and smallest readings. The range is quickly and easily found and can be a useful pointer to the overall spread of data. However, as the only values taken are the two at the extremes, it may give a very misleading idea of the general pattern of scatter. Some extremes can be very extreme indeed, dwarfs and giants for example, in relation to most people's height.

INTER-QUARTILE RANGE

A simple way of discounting the effect of extremes is to exclude the 25% of readings at the tail-ends of the range of data. This is a sensible though rather arbitrary thing to do, if we know that to include these readings would otherwise give a very deceptive view of the distribution. The inter-quartile range covers the range to be found within the middle 50% of the readings.

To express this in more statistical language, it will be remembered that with all items arranged in order of magnitude, the reading corresponding to the middle item is called the median. The values corresponding to the one-quarter and three-quarters positions are known as the lower and upper quartile values. The difference between them is called the inter-quartile range. Half of it is the quartile deviation, i.e. the average amount by which the two quartiles differ from the median.

MEAN DEVIATION

One of the clearest indications of the extent of scatter around an average, which uses all the readings, can be gained by measuring how much individual readings differ – or deviate – from the average. When data are closely bunched this figure will be small, when widely dispersed very much larger.

To measure the Mean Deviation, deviations from the average are totalled ignoring signs, and the average of these deviations calculated. Here is a simple example (*d* stands for deviation and *pmh* means per man-hour):

	Output: units pmh On flat rate	*d*	*d²*	*Output: units pmh On incentive scheme*	*d*	*d²*
A	97	3	9	77	23	529
B	98	2	4	81	19	361
C	100	0	0	100	0	0
D	101	1	1	110	10	100
E	104	4	16	132	32	1,024
		10	30		84	2,014

Arithmetic Mean = 100
Mean Deviation = 10/5 = 2
Standard Deviation = $\sqrt{(30/5)} = 2 \cdot 5$ (approx.)

Arithmetic Mean = 100
Mean Deviation = 84/5 = 16·8
Standard Deviation = $\sqrt{(2,014/5)} =$ 20·0 (approx.)

Ignore for a moment the columns headed d^2. The five workers A to E when employed on a flat-rate basis have an hourly output varying from 97 units to 104. Average output is 100 units per hour. Only one man in fact does 100 units, two do a few less, two a few more. The deviation column records how each man's output differs from the average. To know the average variation – or mean deviation – add up the

27

individual deviations from the average of 100, ignoring signs, and divide by five. In this instance, it is 2 units.

The men are now put on an incentive scheme. One result is a much greater variation of individual output. Average output happens to remain the same, but two men have rather given up the challenge. Their output has slumped. One man has responded magnificently, raising his output by almost a third. The much greater variability of individual output shows up in the mean deviation, which is now 17 units, compared with 2 before.

The calculation of the mean deviation as a measure of scatter is straightforward. It has the advantage of using all the data. However, it has a statistical limitation that we cannot build upon it in as rich a way as happens to be possible if we use an alternative measure known as the standard deviation.

STANDARD DEVIATION

The standard deviation, a variant of the mean deviation calculation, is a key measure in statistical analysis.

When calculating the mean deviation, the deviations were totalled ignoring the signs. An alternative device is to square all the deviations, divide by the number of observations, and then take the square root. In practice, some refinement of this procedure is necessary and there are many adaptations to cope with more complex data, but the essential notion for calculating the standard deviation remains the same.

In the example, comparing the standard with the mean deviation, it will be seen that they are roughly of the same order: mean deviations of 2 and 17 units, standard deviations of 2·5 and 20 units. However, as has been said, it is the standard deviation which proves the more powerful tool.

As a measure of dispersion, the standard deviation is of the greatest value for further statistical analysis, particularly in applications of the theory of probability. Even a modest acquaintance with the term and what it represents can open up an important area of understanding in statistical analysis. This will become evident later from a consideration of certain other statistical notions and measures, which hinge upon the standard deviation.

(An aside on terminology. The square of the standard deviation is known as the 'variance', a term with a strictly technical meaning in statistics. Accountants too use the term 'variance' but for a very different notion.)

To summarize what we have been saying in this section. A central value, coupled with a good measure of the scatter of observations around this value, can describe a frequency distribution in a form of statistical shorthand, which then has the additional attribute that it lends itself to more extended statistical analysis and application. For those who have to make policies and take decisions, this extra statistical power opens up a whole range of measurements and analytical aids – some quite simple, some complicated – which are theirs for the using, provided they understand something of their general nature and provided the statisticians can explain their particular skills and techniques in reasonably practical terms.

Chapter 3

Taking Samples

'If we postulate that within un-, sub- or supernatural forces *the probability is* that the law of probability will not operate as a factor, then we must accept that the probability of the *first* part will not operate as a factor, in which case the law of probability *will* operate as a factor with un-, sub- or supernatural forces. And since it obviously hasn't been doing so, we can take it that we are not held within un-, sub- or supernatural forces after all; in all probability, that is.'

Tom Stoppard*

*Tom Stoppard, *Rosencrantz and Guildenstern Are Dead.* Faber and Faber, 1967

3

Taking Samples

In ordinary life we rely greatly on sampling. The methods we use may be somewhat haphazard and the outcome not always satisfactory, but we take it for granted that many decisions can be effectively made only on the basis of the impressions we gain from a sample, whether we are buying grapes from the barrow boy, selecting from the tailor's pattern-book or choosing a paint from a colour card.

Sampling applied to managerial problems, however, must be done systematically and with a conscious concern for skill and economy. Preferably also, methods of sampling would be welcomed which make it possible to evaluate the level of dependability achieved. Put rather differently, a measure of the risk involved in taking a decision on the strength of sample results would be very useful. Sampling is bound to involve risk. It is an advantage of statistical sampling that these risks can be calculated. One can then take a view on the order of risk one is prepared to run in given circumstances.

This point about risk, far from being a weakness, is one of the strengths of sampling, particularly once we have freed ourselves of the notion that 100% surveys or checks are risk-free. There are a number of reasons why in situations where one is dependent upon human judgement, even assuming a 100% coverage practical and economic, the results could be less reliable than might be thought. The fallibility of human inspection on production lines is well documented. There is, for example, the disconcerting observation that even with the

33

most carefully calibrated instruments, a tendency persists to see the reading one expects rather than the reading actually registered on the instrument.

On the variability of individual judgement and interpretation: 'One of the most striking demonstrations of this arose out of an inquiry into the most suitable radiographic technique for making routine chest examinations (Birkelo *et al.*, 1947). Five chest experts examined 1,256 radiographs and classified each as positive or negative for tuberculosis. There were considerable discrepancies between the experts' opinions, for instance one expert picked out fifty-six positives and another one hundred. When the experts reassessed the same films two months later there were discrepancies between the first and second opinions: thus one reader picked out fifty-nine films as positive on his first reading and seventy-eight in the second. Of his first-reading positives, seven per cent were returned negative on second reading, and twenty-nine per cent of his second-reading positives had been returned negative on first reading.'*

Once one moves away from the illusion that a 100% check is by definition foolproof, the case for sampling becomes very strong. Moreover, in many instances a 100% check would simply not be practical, as when the check involves life testing or destructive tests. One can hardly run each car engine or television tube the equivalent of its rated lifetime before sale, nor is it feasible to test every single match or bullet without destroying it. Finally, there is the test of practice. The body of experience in the application of sampling techniques to quite diverse problems is now very substantial.

Business applications of statistical sampling are fairly recent in their development. Systematic methods of sampling first evolved in such areas as the study of social problems. Sample surveys made it possible to discuss the problem of poverty, for example, in measurable terms.

Rowntree, a pioneer of such work, undertook his first social survey of York in 1899, his second in 1936. Both were based

* M. L. J. Abercrombie, *The Anatomy of Human Judgement*, page 90. Hutchinson, 1960.

on a house-to-house investigation covering almost every working-class family in York. In the case of the second, Rowntree used the information to test the kind of error that would have resulted, if instead of surveying the whole of the working-class population on an individual household basis he had depended upon samples. He incorporated the outcome of this examination in a supplementary chapter to *Poverty and Progress*. The following table is typical of his findings:*

Percentage of income spent on rent

Income class	Complete survey	1 *in* 10	1 *in* 20	1 *in* 30	1 *in* 40	1 *in* 50
			Sample surveys			
A	26·5	26·6	25·9	27·0	28·3	27·1
	(1,748)	(175)	(86)	(56)	(45)	(34)
B	22·7	22·9	23·5	23·3	22·3	22·6
	(2,477)	(289)	(160)	(97)	(81)	(67)
C	19·8	18·1	17·2	18·3	17·2	18·0
	(2,514)	(238)	(98)	(93)	(46)	(46)
D	15·8	16·0	14·4	15·8	17·1	16·9
	(1,676)	(185)	(83)	(66)	(46)	(35)
E	11·3	11·0	10·1	10·7	11·2	11·5
	(3,740)	(414)	(203)	(132)	(101)	(77)

Percentages are shown; actual numbers are in brackets.
Families classified by weekly income, class A being the lowest income group.

Even at first glance it will be seen that the results which would have been obtained from using samples are of the same order as those derived from the complete survey. When a third survey was planned in 1950† it was decided that substantially accurate results would be obtained on the basis of a sample of 1 in 10. In the event, a sample of 1 in 9 was used, but as will be mentioned later, the confidence we can have in a sample does not depend simply on the proportion of the total population sampled.

* B. Seebohm Rowntree, *Poverty and Progress: A Second Social Survey of York*, page 489. Longmans, Green, 1941. (For further explanation, see the Supplementary chapter, 'An examination of the reliability of social statistics based on the sampling method', pages 478–92.)
† B. S. Rowntree & G. R. Lavers, *Poverty and the Welfare State*. Longmans, Green, 1951.

The business world has drawn on such methods to help deal with the growing complexities of marketing and production and has developed them further. The last thirty years have witnessed the widespread application of sampling techniques to market surveys and research. In the control of quality, pioneer work was done in the 1920s, though it needed the 1939–45 war to spread the idea. Now statistical sampling has moved rapidly into other fields.

THE BASIS OF RELIABLE SAMPLING

The dependability of sampling is clearly governed by the way the sampling is done.

Statistical sampling starts from the fact that its focus is upon groups, populations, aggregates, not upon individuals. It is concerned with what happens on the average, not on what happens in any specific case, and it needs to be remembered that what holds good for a group need not necessarily hold good for an individual within the group taken in isolation.

Here one might pay tribute to one of the great martyrs in the cause of statistics, Girolamo Cardano, an Italian mathematician of the sixteenth century, also a dabbler in astrology. It is said that on the basis of probability theory he forecast the likelihood of his own death on a particular date. (This is the kind of calculation which lies behind every insurance policy.) Rather than run the risk of having his forecast proved wrong – as he mistakenly reasoned – he starved himself over the period prior to the fated day. Perhaps his astrology took over at the wrong moment, leading him to conclude wrongly that what is observed to hold good in general must necessarily apply in a given instance.

Given its focus upon the pattern of behaviour of groups, sampling then exploits the knowledge that there is a tendency for variations in a group to balance out as the group grows larger, and for a small group chosen at random from a larger group to show characteristics that are typical of the larger group.

As an illustration of the tendency for variations in a group to

36

Taking Samples

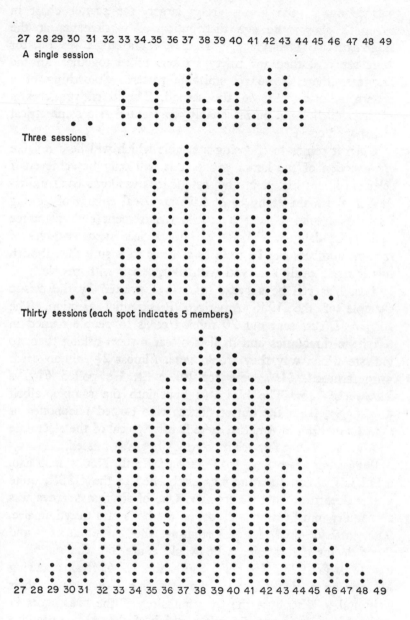

Figure 2.   Age distribution: scatter chart

37

balance out as the group grows larger, the scatter chart in Figure 2 shows the age distribution of participants at the Administrative Staff College, Henley, for a single session, for three sessions, then for thirty sessions taken together. In the first case there is no recognizable pattern. Something of a pattern is discernible in the second. The third produces a picture which could be reasonably represented as a symmetrical bell-shaped curve.

When it comes to choosing a sample which will be a reliable cross-section of the larger group, this will only be achieved if selection is not biased in any way, if it is random. Each individual in the main group must have an equal chance of getting into the sample. This is a crucial requirement and guarantee against the dangers of bias. Various devices such as tables of random numbers can be used as controls to ensure this, though this is more easily achieved with things than with people.

That bias can be a great snare is illustrated by the classic example of the 1936 American Presidential election. The *Literary Digest* sent out 20 million cards to people named in telephone directories and in lists of car owners asking them to indicate which way they would vote. Almost $2\frac{1}{2}$ million cards were returned. 41% were for Roosevelt. He polled 61% a few weeks later. The bias was built into the sample, albeit unwittingly, from the outset. Those who owned telephones or ran cars in 1936 in America were hardly typical of the electorate as a whole. The point seems obvious but was missed.

But lessons often have to be learned afresh. Not so long ago, a Public Opinion Institute was established in the USSR, quite a new departure for that country. One of its first surveys was an enquiry into living standards, on the basis of a questionnaire. The method used for distributing the questionnaire – and therefore for determining who would come into the sample – was to arrange for the train conductors of 65 trains leaving Moscow on a given date to choose a single carriage in their train and to distribute the questionnaires to the passengers in that carriage. It is not difficult to think of the many questions one could ask about how far this was likely to constitute a representative sample of the population. However, it must be remembered that the idea of social surveys of this kind was novel and that there had been little reason or opportunity for people to become versed in the appropriate sampling methodology. The USSR is not short of people with the ability to acquire analytical skills and techniques, once they are alerted to the problem and to the body of knowledge and experience available elsewhere.

Given a sound sampling technique, the larger the sample the more accurate the results, but contrary to what one might expect, gains in accuracy are far from proportional to increases in sample size. (To understand why this is so would entail going into the mathematics of sampling.)

Important factors which would influence the size of the sample are the characteristics and amount of variability likely to be at work in the population under study and the amount of detail and quality of information one is seeking.

The approach adopted in carrying out the National Food Survey illustrates the design of a sample intended to provide a coverage of the whole of Great Britain.

'In order to make the sample representative it is necessary to cover households of different family composition and social class, and to take into account their distribution by region and type of area.

'The sample was selected by a three-stage sampling scheme, involving at the first stage the selection of 50 parliamentary constituencies in Great Britain. The second stage consisted of the selection of a number of polling districts within these constituencies, and the third stage the selection of a number of households within each of the polling districts chosen at the second stage.

'In selecting the constituencies it was decided to exclude the six in the crofting counties of Scotland because the cost of sending fieldworkers to their widely scattered households would have been prohibitive. These counties contain only 0.6% of the population of Great Britain.

'The third stage of sampling consisted of the selection with equal probability of approximately 17,000 addresses from the electoral registers of the selected polling districts. About 340 addresses were selected from each constituency on the basis of 85 each quarter. Because of failure to reach the housewife for reasons such as the illness of the interviewer, about 16,600 households only were effectively covered, a contact rate of about 98%, and from this number, a response rate of almost 54% resulted in a final sample of 8,931 households from 831 polling districts in which households were visited.'*

Random – or 'probability' – sampling serves to avoid bias. It has a further advantage. It makes it possible to calculate the odds on a sample measure holding good for the population from which it was drawn, in other words to evaluate the order of risk one would be running in assuming that what has been learned from the sample holds true for the whole.

Sample reliability can be measured because where chance causes alone operate the resulting frequency distribution takes on a characteristic pattern known as the 'normal' curve, and the mathematics of this curve are well understood. This knowledge becomes a powerful tool in sampling techniques.

While it is not essential for the layman to know about the mathematics of the 'normal' curve, some acquaintance with its general characteristics is desirable.

THE 'NORMAL' DISTRIBUTION CURVE

The point needs to be made first that there are many probability distributions differing from each other in their pattern of scatter, but a quite limited number of families of distribution curves have been found in practice to be applicable to data in diverse fields.

The 'normal' curve is just one of a number of mathematically 'ideal' curves. It happens to describe accurately the distribution pattern of data relating to many natural, economic, and social phenomena. There is a temptation to assume rather too readily that data are 'normally' distributed. Statistical tests can be applied to establish if a particular distribution does reasonably conform to the 'normal'. But even though the original data may

---

* *Domestic Food Consumption and Expenditure:* 1957. (Extracts from Appendix A: Composition of the Sample.) HMSO.

not be 'normally' distributed, we are helped by the fact that the measures computed from a number of samples themselves tend to be 'normally' distributed.

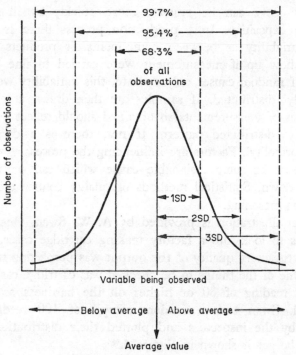

Figure 3. The 'normal' curve

As Figure 3 shows, the 'normal' curve is symmetrical and bell-shaped. Because of its symmetry, the arithmetic mean, median and mode coincide. Each particular curve can be defined in terms of its arithmetic mean (A.M.) and standard deviation (S.D.). The points of inflection – where curvature changes direction – arc 1 S.D. from the A.M. From the mathematics of the curve we know that:

68·3% of the total observations are within ± 1 S.D.
95·4% of the total observations are within ± 2 S.D.
99·7% of the total observations are within ± 3 S.D.

i.e. almost all observations lie within a range of 6 S.D.

41

Put another way, with a random distribution, the probability of an item falling outside 1 S.D. is 1 in 3, outside 2 S.D. is 1 in 20, outside 3 S.D. is 3 in a thousand.

The general idea which underlies the application of the 'normal' curve can be grasped more quickly by illustration than by exposition. In a production process there is always some variability as between components or products, which would show up if measurement were carried to fine enough limits. If random causes only operate, this variability would be 'normally' distributed. If samples are then drawn at random, the series of measurements so obtained should reflect a similar 'normally' distributed pattern. If not, there is a divergence from 'normality'. Factors are influencing the process, for which there must be some assignable cause which can perhaps be tracked down. Statistical methods of quality control are based upon this reasoning.

A neat illustration is provided by A. W. Swan. One of his jobs was to look into a factory making cartridge cases, where the unsatisfactory quality of the output was perplexing management. One of the things to be checked was cartridge case hardness. A reading of 80 or higher on the hardness scale was required. Swan took a random sample of 100 readings as logged by the inspectors and plotted their distribution. The picture he got is shown in Figure 4.*

Since the sample readings were taken at random, unless bias had been introduced into the inspection process, the distribution of the sample readings around the average should also have been random, and the picture should have approximated to the 'normal' curve. From the picture obtained, it was evident that some factor was at work distorting the inspection process and introducing bias. The dip in the curve just below the 80 mark was quite unnatural. Given this evidence, it was not very difficult to make further checks to uncover where this 'bias' was coming from and the case for checking the practices of the inspectors did not then rest solely on surmise or suspicion.

* A. W. Swan, 'Operational research today and tomorrow', *O.R. Quarterly*, December 1958, page 279.

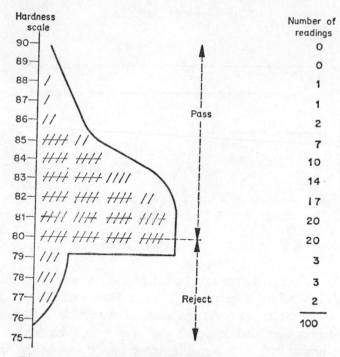

Figure 4.   Distribution of hardness of cartridge cases

THE QUALITY CONTROL CHART

The quality control chart exploits this way of looking at a production process. A typical quality control chart defines the predetermined pattern to which the output of a process would conform, assuming nothing upsets smooth running. Measurements would be 'normally' distributed. The vast bulk of output would fall around the specified dimensions within agreed tolerances. Any variability would be natural to the process and solely due to chance. If chance causes alone are operating, 19 readings in 20 would fall within the limits of ± twice the standard deviation, only 3 in 1,000 would fall outside the limits of ± three times the standard deviation.

If random samples are taken, the resulting pattern should also be 'normally' distributed. The pattern that emerges on the control chart through taking periodic sample checks while the production process is under way signals how the process is behaving. (Figure 5).

43

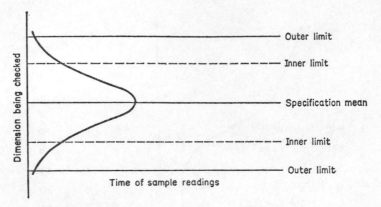

Figure 5. A quality control chart and the 'normal' curve

The tolerance limits can be set in such a way that trouble shows up before it gets out of hand. For example, the outer limits may be set to cover the range of acceptable output, with the expectation that most of the output will be within the inner limits, with variation distributed evenly around the specified dimension. If an undue number of sample readings fall between the inner and outer limits or concentrate on one side or the other of the specification mean, all the output would still be acceptable but such indications would serve as warning that the process threatens to go off balance. The causes of this disturbance to the process can then be traced before they do damage.

There are many situations where one is operating a process. One is concerned to monitor how the process is running, to make sure that it goes on running within the limits of prescribed conditions and to judge when and how much of a correction may be needed to regulate it and keep it under control. A great deal can be learned about the way a process is behaving from a sequence of measurements and we can extend the use of sampling and quality control charts to provide a picture of how the process is behaving, measured against the conditions specified for it. The various small changes can be cumulated and graphed in such a way as to give quite sensitive indications of when changes from specification are occurring, up to the point that a need to intervene is signalled and some correction applied. 'Cumulative sum techniques', as they are known, offer a valuable way of monitoring and controlling processes, through the use they make of statistical and graphical methods of quality and process control.

The approach to operating a process in a way which generates by design a continuous, planned flow of information on how to improve the output and regulation of the process has come to be known as 'evolutionary operation'. It is natural that its main application should have first been in the chemical industry. It is an approach in which statistical design plays an important part, though the operating procedures and calculations involved are usually quite simple.

QUOTA SAMPLING

While random sampling may be statistically the best, in market survey and similar work it can prove awkward and relatively expensive. For example, once the individuals are selected no substitution can be allowed. Quota sampling, possibly allied with limited random sampling, is often adopted as an alternative. A model is set up to match the population as a whole and the investigator is left free to select those individuals who will in his judgement fit the quota. As a method it can be cheap and informative, even though it lacks the rigour and the analytical advantages of random sampling.

(The characteristics of quota as opposed to random sampling have been very effectively stated in a study of *Readership Surveys* in *Readings in Market Research*, pages 127–8, British Market Research Bureau Ltd., 1956.)

Recent developments in sampling techniques whereby the probabilities of selection are intentionally designed to be unequal may give the impression that they are a departure from random sampling. This is not so. Probabilities of selection can be manipulated without jeopardy to the sampling process itself, and important economies are to be made by so doing.

There is no dearth of illustrations of the application of sampling to practical problems. To cite two classics. £200,000 were saved because it was established by sample survey that only a third of the 7,000,000 people entitled to some 20,000,000 war medals would in fact claim them. The Post Office, having established that telephone directories played only a limited part in finding numbers, estimated that if new directories were not issued for a year they would hardly be missed and that an increase of not more than 1·3% in the use of the enquiries services would result. £270,000 were saved by suspending the issue of new directories for a year. In the first eight months, of a total of 250,000 subscribers, only 56 enquired why there had been no new issue.

Statistical sampling has been successfully applied in very varied fields. Even where a 100% coverage is feasible, a good case can be argued for a well-conceived sampling scheme as giving dependable results, quickly and cheaply.

Some people will always be uneasy about accepting the idea that knowledge derived from a sample is really satisfactory. Perhaps the most persuasive influence would be to make them more consciously aware of the way in which they sample every day of their lives, in judging a product from the sample seen in a shop window or a glass of beer from the first sip.

It will be evident that the design of a sampling scheme requires a combination of the skills of the statistician with the judgement and decisions of the manager. Since it is rare to find the two combined in the one person, this implies a considerable measure of mutual understanding and collaboration. While the onus must be on the statistician to explain his ideas plainly, it helps if the manager has an acquaintance with the thinking which guides the statistician in applying his expertise.

Chapter 4

Analysing Trends

'Ah, but my Computations, People say,
Reduced the Year to better reckoning? – Nay,
'Twas only striking from the Calendar
Unborn Tomorrow, and dead Yesterday.'

Omar Khayyam*

*Omar Khayyam, *Quatrain*, 4th edition, in
Persian Poems, ed. A. J. Arberry. Everyman's
Library, 1954, page 25

ROCKHURST COLLEGE LIBRARY

4

Analysing Trends

If forward thinking is to be realistically based, we must first extract such guidance as the past can offer. On the assumption that the future follows the past with a measure of continuity, it then becomes important to distinguish factors likely to have a bearing on the future from those which are transitory.

Underlying any such attempt is an attitude towards forecasting. Some say we cannot forecast, others that we have no choice. The view taken here is that the only argument of importance is about how far to make the process explicit and systematic. I assume and forecast that the sun will rise tomorrow. I know no way of guaranteeing this. It is simply an assumption. A decision not to work on this assumption would have quite an influence on the way I spend the next hours of my life.

Similarly, taking a more serious example, one hopes that government policy is being shaped on the assumption – or forecast – that some 7,000 people are likely to be killed in fatal road accidents within the next twelve months. There can be argument about whether it will be nearer 6,000 or 8,000 but few would accept the view that since we cannot forecast the precise number, therefore we have no basis for a sensible road safety policy. That would be tantamount to saying that the number of fatal accidents could be anything. All sorts of eventualities are possible, but the point is that some are much more likely than others.

To take a further example, one trusts that those discussing the implications of the world's population explosion accept that unless something is done to prevent it, today's world population of around 3,600 millions is likely to grow to some 6,000 millions in thirty years' time.

Such an example leads to a general comment about the nature of forecasting. Measures might be taken to deal with the threatened population increase by preventing it, through family planning and similar policies. Assuming this were done and that it proved effective, world population thirty years hence may be, say, 4,000 millions rather than 6,000 millions. No doubt someone would then point to the unreliability of the forecasters, who had said that the population would be 6,000 millions. Much of the forecaster's work is and ought to be in a sense self-defeating. In many situations it ought to lead to policies which forestall the situation forecast. Perhaps this is why forecasters need to be thick-skinned or recluses.

The attempt to forecast has further justification. It is one of the most effective ways of testing how much we really understand our business and the forces which are at work on it. In a sense it also expresses how far we feel in control of events and how far at the mercy of them. Thereby, it serves to check on our understanding, on the assumptions we make and the attitudes we hold towards the future, as well as on the hopes we have of using experience and knowledge to gain some mastery over the future.

'TIME SERIES' ANALYSIS

There are various ways of classifying the movement of events. In studying problems of an economic nature, it has proved useful to segregate what are regarded as longer-term trends from shorter-term movements and seasonal effects.

Despite the distortions of abnormal events, certain LONG-TERM TRENDS can be detected in line with what might be termed the broad sweep of historical development. The forces at work can be seen, for example, in the steady reduction of hours of work, in the ageing of our population, in the contest between traditional textiles and man-made fibres, television and the cinema, private cars and public transport.

In the history of business activity, periods of prosperity and depression have tended to alternate. There may be dispute as to how far these FLUCTUATIONS are correctly described as cyclical, but none about the fact that there are these ups and downs. Whereas pre-war they appeared to spread over a period of seven to eleven years and were severe in amplitude, since the war they have been much less severe and have been nearer to four years in their spread. Understanding of the causes of these fluctuations is improving. It has been tested in recent years through governmental and international efforts to avoid the serious depressions of the inter-war period, more generally to induce greater stability in economic affairs, while not impairing the tendencies which make for healthy expansion.

In many activities SEASONAL UPS AND DOWNS occur. These may need no explanation, but the problems of planning and running a business which is vulnerable to seasonal fluctuations are such that it becomes important to assess the extent of the seasonal impact and how best to legislate for it.

Finally, there will always be factors whose working we do not understand, which leave unexplained a RESIDUAL element in any trend. This element may be large or small. If large, the main weight of study would then be directed to reducing it.

'Time series' analysis endeavours to pick out the various elements which make up the overall movement. This can be tricky. Some boost and some counteract one another. Short-term movements can be quite deceptive. As misinterpretation could be disastrous for policy decisions, it becomes vital to strike a right balance between the longer-term perspective and the short-term view, an easy thing to say but much more difficult to do.

Short term movements show up very clearly in sales of consumer goods. Figure 6 illustrates some of the fluctuations and trends for retail sales – which cover about half of consumer spending – over a period of five years. Comparison is made easier by the trend lines superimposed on the short-term movements. In terms of current prices, retail sales in total give the appearance of having moved sharply upwards. Revalued at 1966 prices, however, the overall rise is much less startling.

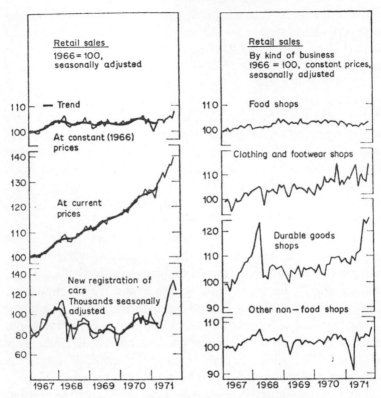

Figure 6.   Retail sales: Great Britain
(From *Economic Trends*, HMSO, December 1971, page 11.)

Turning to specific markets, there is not much doubt about the volatile character of sales of new cars, which stands out even on the trend line. For the other categories of consumer goods charted, we can see the numerous short-term movements, but each market has its own trends and quirks and there are important contrasts between them. Compare clothing and footwear, for example, with food.

Long-term decisions cannot always be insulated from the effects of short-term changes. The car industry, as an example, has to fight a hard battle between the shorter-term restrictions periodically imposed upon it by the buffetings of Government policy and its own conviction that it is in an expanding market judged over decades.

Analysing Trends

GRAPHING

Quite simple graphs can often highlight features of a trend – provided the rules of good graphing are observed.

Introductory text-books usually have a section on this and there is no need to go over the same ground here. An illustration will suffice. Take the one set of figures, say of sales which have risen consistently over a period of months. The Sales Manager might be tempted to describe this as business soaring. To give adequate expression to his expansive attitude, his graph would have a wide vertical but a narrow horizontal scale and the resulting line would indeed soar. The Accountant may be more restrained. While not wishing to appear discouraging, he may nevertheless consider sales to be growing not all that fast. His graph would have a narrower vertical scale and wider horizontal scale and his line would creep up. The Managing Director, ever conscious of how competitors are doing, may know that even though his own Company's sales are rising, they are expanding less fast than those of his competitors. He would use yet another picture, one which shows that his Company's share of the total market is declining. Most graphs are designed as pictures to support a case or an argument. First comes the case to be argued, then the picture suitably dressed-up. One can only advocate that the dressing-up be done skilfully and that at the same time people be alerted to the dangers of being unduly impressed by graphs and charts until they have been carefully scrutinized.

LOGARITHMIC CHARTS

The logarithmic scale is particularly valuable for comparing relative movements, which is what managers usually want to do with figures. It is much easier to use than appears to the uninitiated.

Consider a business which might grow in the following way:

	Turnover £	Cost of Materials £
1970	70,000	30,000
1971	400,000	160,000
1972	1,200,000	500,000
1973	1,800,000	750,000
1974	4,100,000	1,400,000

To bring out the growth of the business, the figures might be plotted in the ordinary way on a natural scale graph. Alternatively, the logarithm of each value can be graphed. Logarithmic graph paper does this for us. It automatically plots figures in their logarithmic equivalents. (*See* Figure 7.)

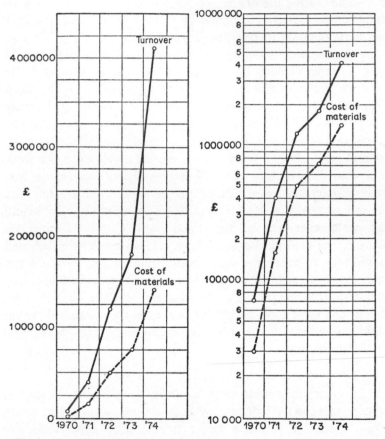

Figure 7.   Natural scale and logarithmic scale graphing compared

How do the two graphs compare? Take the 1973 turnover and cost of materials which, compared with 1972, went up by £600,000 and £250,000 respectively. The absolute figures are not particularly informative. There is little that can be usefully said about either of these figures in isolation. Relatively, both went up by the same proportion – 50% – and it is this relative

54

movement which matters most, since it indicates whether the increase in turnover has been achieved at reasonable cost. Plotted on natural scale, the graph shows a sharper angle of rise for turnover than for material cost. The same figures plotted on a logarithmic scale give a rise at the same angle.

It is a feature of a logarithmic graph that SLOPE reflects RELATIVE RATE OF CHANGE. A COMPARISON OF SLOPES consequently gives an immediate comparison of RELATIVE MOVEMENT. If we plot sets of figures and they have similar slopes, we can immediately say that they are rising or falling at the same rate. It is such comparisons which are the essence of management discussion on the movement and progress of a business, and logarithmic charting provides a useful tool in making them.

Separate logarithmic graphs relating to DATA IN DIF- FERENT UNITS can be brought together easily for purposes of comparison. Even though the scales are quite independent, the slopes of the curves are directly comparable. Consider the following additional data:

	Output (units)	Absenteeism (days)	Plant utilization (%)
1970	500	400	45
1971	4,100	1,700	93
1972	14,400	9,200	96
1973	19,000	7,100	82
1974	46,000	7,900	89

By combining several logarithmic scale cycles – or decks as they are called – such sets of data can be compared in quite a small compass. With judicious plotting, all the data given can be recorded on a sheet of logarithmic graph paper of quite modest size and quick comparisons drawn. Where one scale only needs to be logarithmic, 'semi-log' graph paper is used, but both can be logarithmic, and 'double-log' – or 'log-log' – paper is available.

In summary, logarithmic graphing copes easily with wide ranges of data and with data in different units. The fact that we can immediately compare relative rates of change makes possible a rapid interpretation of otherwise awkward sets of data.

An aside –

A note for those who do not quite remember their logarithms:

$$100 = 10 \times 10 \qquad\qquad\qquad = 10^2$$
$$1,000 = 10 \times 10 \times 10 \qquad\qquad = 10^3$$
$$10,000 = 10 \times 10 \times 10 \times 10 \qquad = 10^4$$
$$100,000 = 10 \times 10 \times 10 \times 10 \times 10 = 10^5$$

The series 2, 3, 4, and 5 are the logarithmic equivalents of the original figures. Logarithmic tables give values for intermediate numbers.

If printed logarithmic paper is not available it is not a difficult matter to draw up one's own, by plotting on ordinary graph paper the logarithms of the numbers instead of the actual numbers. It is a peculiarity of logarithmic graph paper that there is no zero, and negative values cannot be plotted, but there are ways of getting around these limitations.

MOVING AVERAGES AND MEASURES OF SEASONAL EFFECT

Moving averages and moving totals help to make trends more explicit. No less important, they provide one method of calculating a measure of seasonal effect. It is not enough just to know that people eat more ice cream in hot weather than in cold; a manager wants to know how much more.

The calculation of moving averages and totals is simple. The strangeness is only in the idea. Take the following figures of monthly sales over a period of two years:

			Sales		
			Current	*Moving annual total*	*Cumulative*
1971	J	£	8,000		8
	F		7		15
	M		6		21
	A		9		30
	M		10		40
	J		11		51
	J		12		63
	A		6		69
	S		10		79
	O		13		92
	N		12		104
	D		10	114	114

Sales

		Current	Moving annual total	Cumulative
1972	J	£ 8,000	114	8
	F	6	113	14
	M	5	112	19
	A	9	112	28
	M	9	111	37
	J	9	109	46
	J	14	111	60
	A	8	113	68
	S	7	110	75
	O	10	107	85
	N	11	106	96
	D	10	106	106
1973	J			

Sales for the year running from January to December 1971 totalled £114,000. If habit-ridden, we then start afresh in January 1972 and say to ourselves that for purposes of comparison we have to wait until December before another twelve months of experience become available.

However, if we can emancipate ourselves from the calendar year – or from the financial year for that matter – any period of twelve months irrespective of when it starts and ends can be regarded as a year of experience. February 1971 to the end of January 1972 is also twelve months, as is March 1971 to the end of February 1972, and so on. On this basis, given monthly figures for two calendar years, one can derive a series of moving annual totals representing thirteen 'years' and from them moving averages. Each twelve months automatically includes all the seasons. The moving annual totals thus give a trend series which absorbs the ups and downs within the year.

Printed graph paper makes it easy to plot together the month-by-month figures, the cumulative figures for each calendar year and the moving totals, and in this way to build up a continuous picture showing short-term movements against the longer-term trend.

There are a number of ways of analysing seasonal variation. The following borrowed and slightly adapted example explains one, based on moving average and moving total calculations. It has been derived from Connor & Morrell, *Statistics in Theory and*

57

On Thinking Statistically

Practice, pages 52–5 (Pitman, 1964), to which the reader should refer for a more complete explanation. (*Note*: This method is appropriate mainly where seasonal effects are pronounced.)

Fertilizer Sales (thousand tonnes)

	a	*b*	*c*	*d*	*e*	*f*	*g*	*h*
Year and quarter	Sales	Moving annual total	Sum of two M.A.T.s	Moving average trend (c÷8)	Devn. from trend (a—d)	Average seasonal movement	Sales (adjusted) (a–f)	Residual (g–d)
1970: 1	60					+36	24	
2	65					+ 5	60	
3	20			47	−27	−30	50	+ 3
4	44	189		47	−3	− 8	52	+ 5
1971: 1	62	191	380	47	+15	+36	26	−21
2	58	184	375	49	+ 9	+ 5	53	+ 4
3	28	192	376	52	−24	−30	58	+ 6
4	50	198	390	53	− 3	− 8	58	+ 5
1972: 1	85	221	419	52	+33	+36	49	− 3
2	42	205	426	52	−10	+ 5	37	−15
3	33	210	415	55	−22	−30	63	+ 8
4	44	204	414	63	−19	− 8	52	−11
1973: 1	118	237	441	65	+53	+36	82	+17
2	71	266	503	65	+ 6	+ 5	66	+ 1
3	20	253	519	66	−46	−30	50	−16
4	58	267	520	66	− 8	− 8	66	0
1974: 1	110	259	526	68	+42	+36	74	+ 6
2	83	271	530	68	+15	+ 5	78	+10
3	22	273	544			−30	52	
4	55	270	543			− 8	63	

In this case, it should be first noted, the average of the sum of two moving annual totals is taken. The reason is that if quarterly sales for one year are averaged, the result would correspond to a period in time mid-way between the second and third quarters, which would be inconvenient. If, however, two 'years' – five quarters in calendar terms – are taken together and then averaged, the result would correspond in time to the third quarter and the trend series would then line up conveniently with the original periods.

The resulting series of averages gives a moving average trend. The difference between the trend and any particular quarter measures the deviation of that quarter from the trend.

The Seasonal Movements Table considers each quarter over the years. The average of the deviations provides an estimate of seasonal effect or the average seasonal movement. In this instance, peak sales are in the first quarter of the year.

Seasonal movements

Year	Quarter			
	1	2	3	4
1970			−27	− 3
1971	+15	+ 9	−24	− 3
1972	+33	−10	−22	−19
1973	+53	+ 6	−46	− 8
1974	+42	+15		
Total	+143	+20	−119	−33
Average	+36	+ 5	−30	− 8

Each quarter can then be discounted by the average seasonal movement to give an adjusted sales figure. Taking the first quarter of 1972 as an example, sales at 85,000 tonnes were 33,000 tonnes higher than the moving average trend figure. However, in this quarter the seasonal effect could anyway have been expected to boost sales by 36,000 tonnes over the trend figure. On this calculation there is, therefore, a short fall – a residual – of 3,000 tonnes which needs to be explained.

It is interesting to compare graphically the direction of movement of the trend curve with that suggested by the movement of quarterly sales considered alone. (*See* Figure 8.)

The utility of cyclical analysis perhaps can be best judged by way of an illustration. The shoe trade offers a characteristic example where the uncertainties and the ups-and-downs of the market pose very difficult management problems and where there has been a traditional view on the part of the trader and manufacturer that not much can be done about them.

A statistical analysis of the shoe trade cycle suggested otherwise. Leicester* was able to expose very effectively the interacting nature of decisions and to show how the ordering habits and practices of retailers and manufacturers aggravated the problem. For example, one result of the analysis was to reveal that

* C. S. Leicester, *The Cure of the Shoe Trade Cycle*. Department of Applied Economics, University of Cambridge. Reprint Series No. 278, 1968.

deliveries made to retailers were often on the basis of assessments and calculations made eighteen months before. Leicester was also able to show that the forecasting ability of manufacturers was no better than that of the retailers.

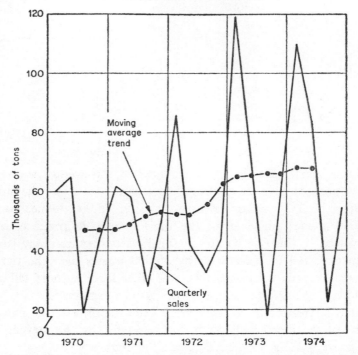

Figure 8.   Fertilizer sales: quarterly movement and the trend

In the shoe trade, the influence of the seasons is, of course, of considerable significance. Analysing deliveries quarter-to-quarter over a three-year period, it is possible to pick out the underlying trend and the seasonal pattern, leaving a measure of the amount of the variation which cannot be explained in either of these ways described as random). To summarize:

The components of deliveries to shoe retailers

£m at con-stant prices	Year 1				Year 2				Year 3			
	I	II	III	IV	I	II	III	IV	I	II	III	IV
Actual	59·1	60·9	60·3	65·1	70·2	65·9	63·6	67·1	74·0	67·8	65·8	65·5
Seasonal pattern	+4·1	−0·8	−2·6	−0·7	+4·1	−0·8	−2·6	−0·7	+4·1	−0·8	−2·6	−0·7
Trend	57·7	60·3	63·3	65·2	66·2	66·4	66·7	67·9	69·1	68·9	67·9	66·8
Random	−2·7	+1·4	−0·4	+0·6	−0·1	+0·3	−0·5	−0·1	+0·8	−0·3	+0·5	−0·6

60

Analysing Trends

Analyses of this kind are not difficult to understand and they can have a very practical value for those who have to make the main policy decisions. Managerially, they can help to get a better grip on the situation. As Leicester argued, at a conference later, 'It is clear that a fashion industry which spends too much time thinking on "what type of shoes" to make and sell, spends too little time thinking of "how many". Failing to get the size of the market right is inefficient trading. Getting these economic trends wrong usually leads to a commercial passing of the buck: in 1961–62, it was the fickleness of the consumer that was blamed. In 1965–66, the industry has only itself to blame.'*

CORRELATION

Undertakings do not operate in a vacuum. Some can relate their destinies quite closely to trends in the economy at large. School building, to an important extent, is a function of changes in the population's age structure, on which there is much useful knowledge. Food manufacture can be planned in the light of known trends of the relations between income and patterns of food consumption. Private car ownership is governed largely by standards of living. How factors such as these are CO-RELATED may be measurable.

The notion of correlation can be illustrated by considering the relationship between output and direct costs, and between prices and sales. An increase in output usually means a more-or-less proportionate increase in direct costs. Output and direct costs move in step with each other, and deviations in one tend to correspond to deviations in the other. The two are closely associated – or correlated. Because they move in the same direction, they would be regarded as 'positively correlated'. There is also a correlation between price and sales. Usually the higher the price, the lower the sales; given a price increase, a fall in sales is generally to be expected. The two move more

* C. S. Leicester, *The Shoe Trade Cycle*. University of Cambridge, Department of Applied Economics. Reprint Series No. 279, 1968.

61

or less in step, but in opposite directions. They would be regarded as 'negatively correlated'.

There are various ways of bringing out how far sets of data appear to be correlated. Initially, a simple tabulation may be quite effective. Graphical methods can be used. A measure of correlation can be calculated.

Clearly, a measure of the closeness of movement would be most useful. However, calculation is complicated by the fact that the data to be compared may be in different units and of very different orders of magnitude. These difficulties can be overcome by expressing each set of data in terms of a scale whose zero is the arithmetic mean of the original data, and where the unit of measurement is the standard deviation of the original distribution. Individual observations for each set of data can then be converted and expressed in terms of the standard deviation of that distribution. In this way distributions are transformed to a common form. They can now be combined to produce an indicator which measures the extent to which they vary together.

This indicator is known as the coefficient of correlation. Its limits are : $+1$ standing for perfect positive correlation, -1 for perfect negative correlation. The basic formula for calculating this coefficient can be readily understood given a grasp of the notion of standard deviation.

Calculation is a statistical operation, which can be applied to any data. A correlation calculation measures a statistical relationship between sets of figures. The results may well lead to conjecture that there is in fact a cause-and-effect relationship. On this the statistical evidence is neutral. Interpretation rests firmly on the judgement of the user. Failure to recognize this has contributed substantially to the very entertaining field of nonsense correlations, but the real difficulty arises where the nonsense is not so obvious. We may not be convinced by the man who infers soda to be the cause of drunkenness, because whether he drinks gin and soda, or whisky and soda or brandy and soda, he still gets drunk, and after all soda is the common factor in each situation, but much the same point was at issue in arguments about the relationship between smoking and cancer.

Analysing Trends

In analysing trends and in seeking to establish relationships which throw light on them, the prime purpose is to serve the future. Some may discount the possibility of bending the future to one's own objectives or the worthwhileness of adapting objectives to a forecast of the future. However, some assessment of what is expected must be made, whether called crystal-gazing, forecasting or forward thinking. As important as the attempt itself is an explicit recognition of the limits of accuracy attaching to such estimates and a clear understanding of this by the user. One good illustration might serve as a model.* *(See* Figure 9)

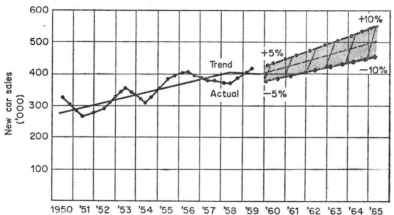

Figure 9.   Canadian automobile market: total annual new car sales 1950-64

Simply and clearly, the point is made that this forecast is estimated to be accurate within 5% for its first year or two. Looking four or five years ahead, an accuracy of the order of 10% should be assumed.

It is up to the analyst to present his conclusions in a form which makes clear the assumptions and the limitations of accuracy which attach to the results. The most sophisticated statistical analysis will never make up for poor judgement when it comes to interpreting statistical findings. One man, who should

* *The Business Quarterly,* Winter 1959, page 228. University of Western Ontario.

know, observes '... the world is full of people who are only too willing to treat a forecast with excessive reverence (or in revulsion dismiss all forecasts as mumbo-jumbo). There are no more dangerous men in Government than those who take figures literally, and this applies both in statistics of the past and to forecasts of the future.'*

Judgement of a high order must also be exercised in defining assumptions on which projections should be based. On this last point, one must not overlook the quite critical difference between projecting the past and predicting the future. The definition of assumptions is not something in which the expert in statistical techniques necessarily has special competence. A fusion of effort is thus needed between the manager with his judgement and intimate knowledge of the business and the expert with his skills of analysis.

* A. Cairncross, 'Economic Forecasting'. *The Economic Journal*, Vol. LXXIX, No. 316, December 1969. (Economic adviser to the Government 1961–64. Head of the Government Economic Service 1964–69.)

Chapter 5

Handling Complexity

'... according to the variety of cases, I can contrive various and endless means of offence and defence'.

Leonardo da Vinci*

Leonardo da Vinci, Phaidon Press Ltd., 1948, page 17, draft of a letter to Ludovic Sforza

5

Handling Complexity

Developments in the application of quantitative method have been rapid and spectacular. A key reason for this is that problems in spheres of interest seemingly remote from each other often can be reduced analytically to a common form. As a result, progress in one field can have rapid repercussions in others.

Queueing theory evolved in the context of designing telephone exchanges. A systematic analytical basis was sought for striking the balance between the extreme situations of subscribers either never experiencing delays or having to wait in long queues. The shape of this problem is not very different from the shape of the problem facing those concerned to determine the right capacity of a bus service, or the right level of stocks to hold of raw materials or finished products. On the one hand there are the costs of a perfect service, on the other the costs implicit in having queues. Theory developed in solving telecommunications problems thus finds ready application in public transport and in business.

It is not easy, however, to link up theoretical advances with the diverse range of possible applications especially to more complex problems. In part this is a problem of jargon. The development of a specialized vocabulary amongst experts to make their own communication more efficient is often indispensable for the advancement of knowledge, but a different language is needed for the uninitiated. To the ordinary motorist, a sign with the number 30 is a speed warning sign. To one systems analyst, it is 'a machine for momentum condensation'. To the chemist, sodium chloride is a compound of sodium and chlorine. To Pierre Teilhard de Chardin, with his vision of the essential unity between the organic and inorganic world, the propensity of molecules to unite was a premonition of love and thus sodium chloride was born out of love of sodium for chlorine. There may be a fascination and much value in the imaginative way words can be used and a new vocabulary fashioned, but for the practising manager this can pose difficulties. As one consequence, the process whereby developments which have a bearing on one's own problems might be spotted is bound to be imperfect, though the effort is none the less worth making.

MODEL BUILDING

The language of model building is spreading, but it can be puzzling to the layman. In ordinary language the word is used in a variety of ways, with the essential notion of representation.

From an analytical point of view, models only become interesting when they can be manipulated, so that one can test the likely effects of possible changes. The familiar pictorial, visual forms of representation are, therefore, of only limited value. They are descriptive rather than explanatory. The photograph of an aircraft gives a quick idea of certain features. It is of no use in determining how the aircraft might behave, assuming one altered the wing span. Organization charts, plans and drawings are in the same category.

The blue-print as a model of a hydraulic system does not enable us to judge what would happen assuming one changed the pressure under which the liquid was being pumped through the system. To find the answer by trying it in practice could prove awkward and expensive. However, developing an analogy between the flow of a liquid and the flow of electricity, an electrical circuit can be designed analogous to the hydraulic system. One can then vary the 'pressure' of electricity, see the effects this has, and by analogy infer what is likely to happen if similar changes were put through the hydraulic system. This kind of model is known as an 'analogue'.

In the case just considered, the flow of electrical energy was used to represent the problem. 'Symbolic' models go a stage further. Symbols, often mathematical, are used to designate properties of the system under study. Because they are abstract and need considerable conceptual skills, such models are the most difficult to establish. But they are the most versatile, since it is a simple matter to change specific properties in the model and work out the likely effects. The familiar formulae of algebra are examples of symbolic models.

Usually, when the highly trained analyst discusses a managerial problem and says that he will proceed to build a model of it, he means that he will write down a few symbols and do some manipulations of them on the back of the traditional envelope. He is model-building, though not in the popular sense.

Just as in ordinary life we know that there may be quite a difference between a model and the real thing, so with model-building in relation to management problems. Model-building involves simplification and therefore a certain loss of realism. Everything depends on how closely the model approximates to reality. Unless there is a close fit, any solution derived from the model cannot be applied to reality, in the expectation that the result derived theoretically will in fact be obtained. Here some allowance must be made for human frailty. Pygmalion became enamoured of the ivory statue he had carved of a beautiful maiden, and there are twentieth-century Pygmalions amongst model-builders.

If managerial problems, by definition, are those which include the complexities of human behaviour, it will be evident that model-building for managerial purposes is still in its early stages. However, it must be acknowledged that impressive developments are taking place in areas assumed until recently to be immune from systematic analysis.

QUANTIFYING PROBLEMS WHICH RESIST MEASUREMENT

The evaluation of research priorities provides a convenient illustration of a problem which, by the very nature of things,

69

involves many unknowns. Yet a decision must somehow be taken on how a research budget is to be allocated. Subjective judgement cannot be avoided, but a rating can be attached to such factors as the importance which individual managers attach to specific research projects, the benefits which they hope might accrue, the likelihood of success, the probable costs and time involved. This process of quantifying subjective notions cannot eliminate the unknowns but it can prove very useful in reducing the boundaries of the problem, clarifying priorities, and coming to a decision. In many managerial problems one must live and grapple with intangibles and complex situations. The effort to quantify them may well illuminate what is otherwise a confusion of haphazard or arbitrary judgments.

NOTIONS OF SELF-REGULATING SYSTEMS

There are many situations where the separation of performance from control creates problems. These are acute in cases where a relatively minor disruption can become amplified and generate more general disruption and instability, if the right action is not taken in time. Feedback theory and cybernetic analysis have led to a better understanding of such problems and of how they might be resolved. Teaching machines, which regulate the selection of problems and the speed with which they are presented to performance, provide one of the clearest illustrations of the application of notions of self-regulation to the more subtle problems of human communication and control.

METHODS OF HANDLING, STORING, AND PROCESSING DATA

The growing use of electronic equipment to process data has impelled the exploration of ways of achieving greater integration in the processing of data for information, control, and planning purposes. These developments have captured the interest of most managers and administrators, even of public imagination. Almost as remarkable has been the progress achieved in the performance of electronic machines, their speed, capacity, and reliability. The influence of these developments on managerial practice is now widely discussed and there is an extensive literature on the subject.

THE MORE ANALYTICAL APPROACH TO FACT-FINDING AND DIAGNOSIS

Increasingly, managerial problems are being diagnosed in a rigorous, scientific manner.

One of the strengths of such an approach is in its freedom to question the obvious or to lay bare the unstated assumption. Babbage, a pioneer in many things, was a good exponent of this. It was he who challenged the logic that the basis of charging for this country's postal system should be variable according to distance. His studies led him to the conclusion that the cost of collecting, stamping, and delivering a letter in this country was so much greater than the transporting cost that there was a strong case for a flat-rate of charges. Sir Rowland Hill introduced the flat-rate penny post not long after.

Consider the example of the Works Manager asked why plant utilization in his factory is only 60%. If typical of his breed, his answer will take the form: Sales keep on changing their minds, Maintenance have no sense of urgency, there are constant hold-ups in material supplies. Anyway, who said 60%? On this last point, at least, he will gain a few weeks of respite, while argument proceeds on what is the right figure.

How might a skilled analyst approach such a situation? Rather than get embroiled in argument about the reliability of past figures, he may take a sample of factory operations by means of a series of snap readings, and thus obtain a realistic measure of utilization quickly and objectively. If the records are in reasonable shape, he might sample them to analyze past delays and their causes. He might be led to question the reasoning behind key controls. For example, the Buying Department may face a rigid limit on the amount of money to be tied up in raw material stocks. This could create problems not only for them but also for the factory, by acting as a restraint on the smooth flow of production, given that orders do not come in perfectly smoothly. Where he finds vital information inadequate and not obtainable by conventional methods, he might well deploy statistical methods of a more sophisticated kind.

In this approach, the independent analyst has certain advantages. He is less likely to be biased or to be inhibited by departmental frontiers. He will have a greater inclination to view things in the round. With such reasons in mind, Professor Kendall has suggested that fact-finding might almost be viewed as a profession in its own right.*

One outcome of such developments is to foster a less fragmented approach to decision making and policy formulation. This is important if only because management problems do not respect departmental frontiers. The growth of operational research has been the most spectacular development of this kind. Cybernetics, 'techno-commercial' analysis, studies of 'socio-technical' systems, equally reflect the quest for more effective co-ordination and integration. Because such terms can cause confusion, those working in the field must help to clarify for the layman the characteristics which make their work distinctive.

The Council of the Operational Research Society give this definition of their subject: 'Operational Research is the attack of modern science on complex problems arising in the direction and management of large systems of men, machines, materials, and money in industry, business, government, and defence. The distinctive approach is to develop a scientific model of the system, incorporating measurements of factors such as chance and risk, with which to predict and compare the outcomes of alternative decisions, strategies or controls. The purpose is to help management determine its policy and actions scientifically.'

Such an approach, coupled with advanced techniques derived from statistical, mathematical, and logical analysis, is being increasingly applied to the solution of practical problems. The following are some pointers.

* M. G. Kendall, 'Modern Statistics in Business and Commerce', *Journal of the Institute of Actuaries*, Vol. 82, 1956.

(A) SEQUENCING AND SCHEDULING

Usually the flow of orders into a business is irregular. This is dealt with partly by using stocks as a buffer, partly by varying the rate of production – which means, in effect, transmitting the irregularity to the shop floor – partly by risking customer goodwill through delays in meeting orders. Mathematical methods of sequencing and scheduling the flow of work have been developed which provide guiding rules, so that informed policies can be established.

(B) QUEUES

Queueing problems, already touched upon, have been the subject of a substantial body of theory. A queue, essentially a lack of balance in time between things coming and going, can arise in many situations, at a bus stop, at the cash desk of a supermarket, at an aerodrome, in a factory maintenance department. The problem is similar, though the setting may be different.

Taking the example of the supermarket, an important decision centres upon the number of cash desks to install. This is not easy to answer. However, if one considers the situation as one where customers arrive at random and the length of servicing time needed for each customer is also randomly distributed, one can then estimate the probability of a queue of given length developing, the average likely time of waiting or the average time during which the cash desks are not being used. The choice can be made in the light of such information, striking the inevitable compromise between having so many cash desks that no customer has to wait and having so few that most customers find themselves wasting a lot of time in a queue.

(C) REPLACEMENT

Problems of replacement policy can be acute where failure or deterioration in a component could have very costly consequences or could seriously lower efficiency. Using estimates of risks of failure and corresponding costs, one can calculate the timing of replacement, so that it takes place on a scheduled and pre-planned basis, which balances risk against cost.

(D) CAPITAL BUDGETING

Capital budgeting depends upon forward estimates. Traditional methods of calculating anticipated returns on projected investments and establishing priorities have tended to take inadequate account of the uncertainties associated with such estimates and the importance of methodically discounting the future. Newer approaches are more rigorous in these respects. They are also more helpful, for example, in assessing how far estimates can be out without overthrowing the soundness of a given investment decision.

(E) PLANNING, PROGRAMMING, AND ALLOCATION

The analysis of trends was considered in Chapter 4 as an essential preliminary to forward planning. Much advanced work has gone into the study of trends and the establishment of bases upon which projections can be founded. Econometric models describe mathematically relationships between economic factors and make it possible to measure and predict the effects of changes within the simplified conditions of the models. The usefulness of such analysis is less in providing specific predictions than in giving a better insight into economic inter-relationships, which are a compound of measurable and non-measurable factors, pin-pointing those assumptions which are most vulnerable and testing the effects of different assumptions.

Programming and allocation techniques take a given objective, such as maximum profits or minimum costs, and determine how one should relate and regulate resources to achieve that objective.

Network analysis constitutes an important advance in methods of planning and controlling complex projects and operations. A plan can be worked out which fully incorporates the necessary sequence of activities, gives realistic estimates of target dates, shows up which are the critical activities and explores the effects of any change or delay in the plan as it is implemented.

(F) CRITERIA OF PERFORMANCE AND EFFICIENCY

Measurement of productivity usually involves a great deal of simplification, and even then one may have to grapple with a significant number of variables. Yet productivity measures have been freely bandied about as if their calculation were straight-forward and their validity indisputable.

Shop floor productivity has also been too readily identified with overall efficiency and there has been a tendency to concentrate on factory operations just because they seem to be fairly easily measured. The performance of sales and administration departments or the effectiveness of advertising are not so easily measurable, but in the context of overall efficiency there are many businesses where they merit far greater study than the factory.

The accurate measurement of performance frequently involves considerable analytical complexity and invites an operational research approach.

Understandably, such developments have succeeded best where problems are well-structured. Large areas of policy-making remain outside this orbit. One might go further. It is difficult to imagine a developing society being short of problems, the complexity and novelty of which demand the managerial skills of taking decisions where judgment must be imposed upon data, information, and the results of analysis, where authority must be exercised and the consequences of decisions responsibly accepted.

An outstanding statistician has warned against expecting too much from mathematical analysis. He was referring to textile technology, but his words are equally pertinent to managerial problems.

'In summing up the achievements of mathematics in textile technology, I would say that they have been essential to progress, but unspectacular. Seldom, if ever, has mathematical analysis led to results that were not previously known through empirical investigation. Very few mathematically based calculations have so far provided a sufficient guide to design or action. This is partly because most textile systems are too complicated to be completely described by such models as mathematicians at present can handle, and partly (perhaps) because empirical investigation has been going on very extensively and for a long time. I think that the main achievement of mathematics in textile technology has been in guiding experimentation, and in helping towards an understanding of what is happening by sorting out the main factors at work and making it possible to assess their importance.'*

One comes back to the opening assumptions and to the suggestion that the manager can best help himself by acquiring a tolerable familiarity with what is going on, learning to recognize the kinds of problem amenable to study and analysis by the specialist, and by being sensitive to the discrimination needed in interpreting and applying solutions.

A further word of encouragement for those congenitally opposed to anything mathematical. In 1660, at the age of twenty-seven, Samuel Pepys was appointed Clerk of the Acts, thereby becoming one of the Principal Officers of the Navy. He applied himself diligently to studying the design of ships and the language of seamanship. This led him to a need to learn how to measure timber. For this 'the mathematiques' were necessary. He arranged for some tuition with a Mr Cooper and so we find him starting his diary for 9 July 1662 with the words 'Up by four o'clock, and at my multiplicacion-table hard. . . .' Managers need not perhaps go that far, but numeracy looks like being one of the pre-conditions for effectively holding posts of responsibility.

* L. H. C. Tippett, 'A Retrospect of Mathematics in Textile Technology', *Journal of the Textile Institute*, Vol. 51, No. 8, August 1960.

Chapter 6

Using the Experts

'You are either a good adviser or a bad one. It has nothing to do with the subject; it's advising you have to be good at!'

George Mikes*

*George Mikes, *Mortal Passion*. Penguin, 1968, page 105

6

Using the Experts

Expertise flourishes more easily than managerial skill. All the pressures are in the direction of each area of expertise evolving into a specialism in its own right, distinct from all the others. Within each specialism there is a sense of professional fraternity amongst practitioners and a concern for the esteem of one's own peer group, which means that the professional often tends to look outwards from the enterprise which happens to be employing him for his status, reputation, and standards. The specialist nature of his work means that his skills are readily transferable to another enterprise or setting. In this sense, by comparison, management is much less professionalized. The skills are more specific to a particular enterprise or industry and a manager's reputation is largely internal to the enterprise he is working in.

There is another distinction to be made which is more fundamental. The manager has to centre his efforts on the particular situation and on the particular managerial problems of his enterprise. He will want to draw upon any knowledge and skills which may help him to understand his problems better and resolve them, but his is a managerial world. He is constantly reaching out beyond a particular functional compartment or skill to the needs of the enterprise considered integrally. A manager in any enterprise of size will be able to call on the experts on its staff and can draw on others from outside, but it is his job to ensure that any use which is made of them is really relevant to the aims and needs of that enterprise and relates to the problems which matter most. The manager who becomes obsessed by any one specialism is likely to be a poor manager.

The expert will want to press his expertise, his skills and techniques. He will seek opportunities for exercising them. Experts are entitled to have the confidence of their expertise and it would be a great pity if they were made to hide their light under a bushel. Not many of them do, because they are often more confident than the average manager in expounding their ideas and may seem to be his intellectual superior because of their mastery of a segment of advanced knowledge.

This is very much so in the case of the statistician or the operational research man. He will certainly have some of his roots in mathematics. There are specialist areas which impinge on management where the quite non-mathematical manager can go a long way in his knowledge and understanding – finance for example, descriptive economics and some of the behavioural sciences – but in statistics he cannot get very far because of the need for mathematics for anything beyond the elementary level. Most managers, therefore, must take much of the statistician's competence on trust and the main test of his ability will be by his effects and results, by whether his work leads to appreciable improvements in the quality of decisions taken and in enterprise performance.

The statistician, as has been contended, has a lot to offer, but he depends on the manager for the opportunities to apply his skills. Hence the need for the manager to get some idea of the statistician's special competence and for the statistician to be led to understand how the manager sees his problems, but one man's normal working vocabulary is another man's jargon. On the principle that it is best to work from the familiar to the less familiar, it is up to the statistician to bend over backwards in explaining himself to management, and if he persists hard enough the chances are that some of his own vocabulary is likely to rub off on the manager during the process.

The difficulty is in making the start. If an enterprise has no statistician on its own books or anyone with that kind of competence, how should it begin? The right answer is probably that it should start quite modestly and not necessarily by recruiting a full-time statistician. The manager must learn what it is like to work with the statistician; in turn the statistician must be given an opportunity to get to know something about the enterprise, its main activities and ways of working. Also, in the early stages, neither would want to get over-committed. To meet this situation, one practical way is for the enterprise to seek out a statistician on the staff of the local college or university. Nowadays, it is often possible to find capable people nearby and usually they are allowed a limited amount of time for consultancy work. Assume that a firm is prepared to risk the annual price of a clerk, this would enable a part-time consultant to put in somewhere between 20 and 30 working days. Spread over a year, this would give him a chance to get to know the essentials of the business, to develop working relationships with managers, and to see where his expertise might bear fruit. It would be surprising if by the end of a year on this basis he had not found ways of giving value for the money. If he does not, the enterprise has not lost too much financially. More importantly, it has not committed itself, in a way which would apply if a man had been appointed full-time. If this approach produces satisfactory results it can be continued on a similar basis or it provides the justification and experience to go ahead with recruiting a full-time statistician.

An alternative would be to use a consultancy firm for one or two particular applications, but a difficulty is that it is not easy to sort out quickly which are likely to be the most worthwhile problems to work on and consultancy firms cannot be expected to wax enthusiastic over very small assignments.

Once an enterprise is satisfied that there is scope for statistical expertise to the extent that a full-time staff is justified, one must be ready for problems of growth, direction, and organization.

Nowadays this area of expertise can take the shape of an operational research unit or a management services department and in addition perhaps the use of an outside consultant in certain specialized areas. This may well be a desirable evolution but it raises problems which it is important to consider and anticipate. Two occasional papers by Ben Aston,* offer a fuller analysis and discussion of these questions.

The reason for the emergence of management services, as Aston points out, is the gap between the needs of an enterprise as perceived by the senior management and its own resources of knowledge and experience available to meet these needs. Thus the services extend the resources and range of skills at the disposal of operating management, but there are conflicting forces at work. After a time, some of the newer notions and techniques offered by these services may become part of the competence of a line manager, and the need for them to be provided as a specialist service may diminish.

On the other hand, this process is slower than the growth in new knowledge and techniques and usually these must be channelled through an individual or unit in the enterprise so that there is a focus for up-to-date thinking. Where the disparity is large, and yet an enterprise is convinced of the value of these newer developments, the result is a rapid expansion in specialist activities relative to the organization as a whole. The aim is to ensure that there will be a rapid and effective exploitation of these services. But it is difficult to move faster than the capacity of managers to understand what these services can give them or their willingness to bring in the specialists. Organizationally, problems and tensions then result. A certain amount of healthy tension is no bad thing, provided energies are being directed to the task in hand. However, all too often these tensions can become a source of bad working relationships and of friction between individuals and departments.

Some of these difficulties may be inherent in the situation. Managers want to feel that they have a grip on things, they will tend to look to the wider issues and to opt for stability. Preferably, they will want change to take place gradually and to be very carefully engineered. By contrast, much of the work of specialists is in the direction of generating change. Often, for

---

* B. R. Aston, *Management Consultancy*, Occasional Paper No. 6, The Administrative Staff College, 1967.

Management Services Techniques and departments. Occasional Paper No. 11, The Administrative Staff College, 1969.

reasons not always of their own making, they may have a fairly narrow and restricted perspective. It is important to understand clearly which of the difficulties are really part and parcel of the situation and which due to clashes of personality or departmental parochialism; the tendency is to turn to the latter explanation too readily.

Aston finds that management services departments may take in the following range of activities: systems analysis and systems engineering, data processing services, operational research, economic analysis and planning, organization planning, and behavioural science applications. In addition there would be the more traditional efficiency techniques, such as work study, and organization and methods.

The unifying characteristics of these very mixed activities are that they are basically in the direction of a more scientific approach to management, that they bring in specialists whom it would be otherwise difficult to locate and use in a conventionally organized enterprise, that the organizational relationship is one of service to line managers, that they are no respecters of departmental boundaries and that usually much greater efficacy is achieved when the various specialists can be brought together in teams, the composition of which is determined by the problems being tackled.

There are, therefore, advantages to setting up a special management services department but it is only one of the ways of internal organization. It may be that specialist posts or small teams could be attached to major operating departments with the central management services office acting primarily as a co-ordinating agency. Or an enterprise may make one of its top executives responsible for management services and for nurturing a receptive working environment for them. Generally speaking, thinking is moving in the direction of a task or project-centred approach, with the talents and resources of the enterprise deployed organizationally in a flexible way.

There are enterprises where the way is open for specialists after some experience to move into operational management and perhaps even come back to management services later.

Such a pattern is feasible in some instances and it is desirable that there should be such possibilities.

Organizational arrangements and devices can facilitate good work but they cannot substitute for the unity and co-operation which stem from a sharing of well-understood primary objectives by managers and by specialist staff. However, objectives may not be clear enough to make unity possible or they may be inconsistent or conflicting. That is when developments, such as management services, which in principle may be desirable, can founder and cause more problems than they solve.

Managers feel that it is they who are finally responsible for earning the daily bread. If subjected to too much pressure and disruption, they are likely to go on the defensive. The specialists may be suspected of having too little concern for the day-to-day operations and for taking too cavalier a view of the problems of adaptation and change. It is for the specialists to see their role clearly. Until recently there has been too little readiness on their part to look critically at the way they go about their work. It would be difficult to fault more than a few statistical applications on technical grounds, but there are all too many examples of failures of effectiveness in practice, because of the way the applications have been handled. Banbury,* a leading operational researcher, finds it necessary to draw attention to the need for the following attributes on the part of an operational research man, in addition to those which are primarily scientific and mathematical: a searching curiosity about the context in which he is working, creative ability and insight, sensitivity to motivation and a feeling for 'values'. For a quantitative analyst, these are highly qualitative attributes – and specialist training does little for them. This is why it is to the evidence of experience and effectiveness that one has to look in order to judge competence.

If it is any consolation, statistical consultancy appears to be no less a problem in the academic world than it is in private

---

* J. Banbury, 'Operational Research and Innovation in Management Methods', *Operational Research Quarterly*, Vol. 19, 1968.

and public sector enterprises. Sprent,* in a paper on this he presented to the Royal Statistical Society, felt it necessary to concentrate on the organizational problems, on the need for the statistician to acquire some knowledge about the field in which he is expected to apply his expertise and on the need for an interplay between theory and practice. He is in no doubt that a statistician's advice must be essentially 'a mixture of art and science'.

Management consultants are usually alert to the more elusive skills which distinguish the good consultant from the mediocre or poor. Many of the larger management consultancy firms have developed a competence in statistics and operational research. There has also been a growth of specialist firms providing a consultancy service in these areas. Since the use of consultants is either an alternative way of drawing upon expertise or a supplement to one's own resources, it is important for the manager to have some knowledge about the services available and the typical situations where they can be profitably brought in. The range of firms and agencies is wide and if guidance is needed it may be useful to refer to the Management Consultants' Association. The British Institute of Management also has an information bureau to help prospective clients to choose management consultants, partly based on confidential user reports. As in any other assignment, the use of consultants for statistical work calls for careful prior thought and for the right kind of working relationship between client and consultant.

The point remains though that for very many enterprises for some time to come, it is the quite basic notions and techniques of statistics which will find ready and most profitable application. At this level the managerial and organizational problems are not severe, if there is a willingness on the part of the manager to give statistical analysis a fair chance and provided he and the statistician recognize that they both share responsibility for achieving effectiveness.

---

* P. Sprent, 'Some Problems of Statistical Consultancy'. Read before the Royal Statistical Society, 12th November 1969.

A Note on Further Reading

The literature of statistics ranges widely; in many areas it is substantial and highly specialized.

While it must be for the individual to judge what will best suit his interests and needs, some suggestions on further reading of a general nature may be found useful.

To keep to the spirit of these notes, the selection has been made on the basis that a short list is to be preferred to a long one and that each work named should have some appeal to the non-statistically inclined reader.

M. Abrams, *Social Surveys and Social Action*, Heinemann, 1951.

A. Battersby, *Mathematics in Management*, Pelican, 1966.

F. N. David, *Games, Gods and Gambling*, Griffin, 1962.

E. Duckworth, *A Guide to Operational Research*, Methuen, (2nd edition) 1962.

B. Edwards, *Sources of Economic and Business Statistics*, Heinemann, 1972.

F. de P. Hanika, *New Thinking in Management*, Heinemann, (2nd edition) 1972.

J. M. Harvey, *Sources of Statistics*, Clive Bingley, 1969.

P. G. Moore, *Statistics and the Manager*, Macdonald, 1966.

W. J. Reichmann, *Use and Abuse of Statistics*, Pelican, 1964.

L. H. C. Tippett, *Statistics*, Oxford University Press, (3rd edition) 1968.

Acknowledgments

TO FIRST EDITION

I am grateful to a number of authors and publishers for permission to reproduce material: The Controller, Her Majesty's Stationery Office, for the extracts from the Crowther Report (in Chapter 1) and the National Food Survey (Chapter 3), also tables in Chapter 1; J. L. Roper and Penguin Books Ltd. for the extract from *Labour Problems in West Africa* (Chapter 2); M. L. J. Abercrombie and Hutchinson & Co. (Publishers) Ltd. for the extract from *The Anatomy of Human Judgement* (Chapter 3); the Executors of the late B. Seebohm Rowntree and Longmans, Green & Co. Ltd. for the table from *Poverty and Progress* (Chapter 3); A. W. Swan and the Editor, *Operational Research Quarterly*, for Figure 4; L. R. Connor and A. J. H. Morrell and Sir Isaac Pitman & Sons Ltd. for the tables from *Statistics in Theory and Practice* (Chapter 4) and Figure 8; R. K. Cowan and the Editor, *The Business Quarterly*, University of Western Ontario, for Figure 9; L. H. C. Tippett and the Editor, *The Journal of the Textile Institute*, for the extract in Chapter 5.

Sources have been acknowledged wherever I have been aware of them, but there may be omissions for which I apologize. I am very conscious of how much one works with borrowed knowledge and ideas picked up over time. Naturally, no blame can attach to others for the way I have used their knowledge.

I would like to thank Mr J. P. Martin-Bates, Principal of

On Thinking Statistically

the Administrative Staff College, for encouraging me to prepare these notes for publication. Mr F. de P. Hanika, Lector, Churchill College, Cambridge, friend and mentor, kindly commented on a draft. Mrs P. Hayes's secretarial qualities could not have been bettered.

M. B., *1963*

TO SECOND EDITION

I wish to thank the authors and publishers who kindly gave me permission to use the following supplementary material in this revised edition: The Controller, Her Majesty's Stationery Office, for three tables in Chapter 2 from the *Annual Abstract of Statistics*; A. F. Sillitoe, Penguin Books Ltd., and the Audit Bureau of Circulations Ltd. for Figure 1; Tom Stoppard and Faber and Faber Ltd. for the extract at the head of Chapter 3; C. S. Leicester and the Editor, *Journal of the British Boot and Shoe Institution*, for a table in Chapter 4 and its accompanying textual extract.

M. B., *1972*

Index

Accuracy of data, 11, 15–16
Allocation techniques, 74
Analogues, 68
Annual Abstract of Statistics, 13
Arithmetic mean, 24, 27, 41
Aslib, 15
Averages, 23–6

Babbage, Charles, 71
Base periods, 21
Behavioural sciences, 83
Board of Trade Journal, 13
Budgeting, 74
Bulletin of Labour Statistics, 14
Business Statistics Office, 13–14

Calculation methods, 21
Capital budgeting, 74
Cardano, Girolamo, 36
Census, 4
Central Statistical Office, 11, 12, 13
Charts, 53–6, 68
   See also Graphs
Classification of data, 21–3
Coefficient of correlation, 62
Comparisons, 16–18
Correlation, 61–2
Cost of living, 18, 20–1

Crowther Report, 3
Cybernetics, 71, 72
Cyclical analysis, 59–61

Data
   quality, 11
   processing, 70–1, 83
Decision-making, 5
Deviation, 27–9
Direct costs, 61
Dispersion, *see* Scatter of data

Econometric models, 74
Economic analysis, 83
Economic fluctuations, 51
Economic Reports and Assessments, 13
Economic Trends, 13
Efficiency measurement, 75–6, 83
Employment and Productivity Gazette, 13
Evolutionary operation, 45
Exports, 19–20

Feedback theory, 70
Fluctuations, 51
Forecasting, 49–50
   limits, 63–4
Frequency distribution, 22

Galton, Sir Francis, 6
Graphs, 53
  See also Charts

Hill, Sir Rowland, 71

Imports, 19–20
Independent Television
  Authority, 24–5
Index numbers, 18–21
International Financial Statistics,
  14
Inter-quartile range, 26–7

Logarithmic charts, 53–6
Logarithms, 56
Long-term trends, 50

Management consultants, 18, 85
Management services units, 81–
  4
Managers
  skills, 79–80
  use of statisticians, 80–5
Manchester Statistical Society, 4
Mean, *see* Arithmetic mean
Mean deviation, 27–8
Median, 24, 27, 41
Mode, 24, 41
Model building, 45, 68–9, 72
Monthly Digest of Statistics, 11,
  12–13
Moving averages, 56–9
Moving totals, 56–9

National Food Survey, 39–40
Network analysis, 75
'Normal' distribution curves, 40–
  3
Numeracy, 3, 76

Operational research, 72–6, 83,
  85
  qualities of specialist staff, 84
  setting up units, 81
Organization and methods, 83
Organization planning, 83
Output, 61

Pepys, Samuel, 76
Percentages, 16–18
Planning, 74
Plato, 3–4
Population statistics, 4
Population trends, 49–50
Probability sampling, *see* Random
  samples
Probability theory, 29
Production processes, 42–5
Productivity measurement, 75–6
Programming, 74
Projections, 64
Public Opinion Institute (USSR),
  38

Quality control, 42, 43–5
Quantifying, 69–70
Quartile deviation, 27
Queues, 73
Queueing theory, 67
Quota sampling, 45–6

Random samples, 38, 40
Range, 26–7
Reliability
  of samples, 36–40
  See also Accuracy of data
Replacement, 73–4
Retail price index, 20
Retail sales, 51–2
Rowntree, B. S., 34–5
Russia, 38–9

Sampling, 21, 33–46, 71
  compared with complete sur-
    veys, 34–5
  for market research, 36
  for social surveys, 34–5
  quota, 45–6
  random, 38, 40
  risks, 33–4
  size of sample, 39
Scatter charts, 37–8
Scatter of data, 26–9
Scheduling, 73
Seasonal movements, 56–9
Self-regulating systems, 70
Sequencing, 73
Shoe trade, 59–61
Sinclair, Sir John, 4
Social surveys, 34–5
Standard deviation, 27, 28–9, 41,
  62
Statistical News, 13
Statisticians, 29
  as consultants, 84–5
  part-time appointments, 81
  relationship with management,
    80–5

Statistics
  applications, 5–6
  economic, 14, 18
  historical background, 4–5
  in management consultancy, 85
  official, 12–13, 14
  population, 4
  social, 14
  standardization, 15
Statists, 4
Symbolic models, 69
Systems, analysis, 71–2, 83

Television advertising, 24–6
Terms of trade, 19–20
'Time series' analysis, 50–2
Trade unions, 23, 24
Trend analysis, 49–64, 74

Variance, 29
Variation, *see* Scatter of data
Vocabulary, 67–8

Weights, 21
Work study, 83

Statistics invade our daily life. They influence our attitudes, our decisions, the way we see our problems. We need to know what they can tell us and what they cannot.

What Mr. Brodie has to say is brief and to the point. His lively and lucid exposition brings out the essentials so often obscured by the longer and more technical books. In the process he dispels the too-common notion that statistics are dull.

Written originally with managers in mind, the book has proved invaluable in practice to other busy people, looking for a succinct layman's introduction to statistics and an explanation of the help they can get from a statistician. Men and women in positions of responsibility, in all walks of life – in business, commerce, the public service, and the professions – share the same need: to think statistically.

It has also found use in universities, technical colleges, and in schools, with students of the professions and of the social sciences, where a non-specialist but intelligent understanding of the subject is required.

This new and revised edition includes changes of some substance, but has preserved the distinctive character which made the first edition so successful that it went through four impressions.